THE BIOLOGY AND CONSERVATION
OF ANIMAL POPULATIONS

THE BIOLOGY
AND CONSERVATION
OF ANIMAL
POPULATIONS

JOHN A. VUCETICH

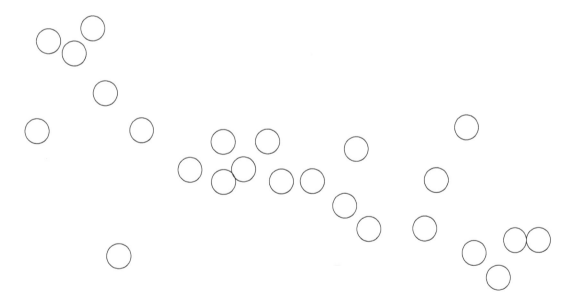

JOHNS HOPKINS UNIVERSITY PRESS | *Baltimore*

© 2024 Johns Hopkins University Press
All rights reserved. Published 2024
Printed in the United States of America on acid-free paper
9 8 7 6 5 4 3 2 1

Johns Hopkins University Press
2715 North Charles Street
Baltimore, Maryland 21218
www.press.jhu.edu

Library of Congress Cataloging-in-Publication Data

Names: Vucetich, John A., author.
Title: The biology and conservation of animal populations /
    John A. Vucetich.
Description: Baltimore : Johns Hopkins University Press, 2024. |
    Includes bibliographical references and index.
Identifiers: LCCN 2023049730 | ISBN 9781421449173 (hardcover) |
    ISBN 9781421449180 (ebook)
Subjects: LCSH: Animal populations—Textbooks. | Conservation
    biology—Textbooks. | Population biology—Mathematics—Textbooks. |
    Animal populations—Mathematics—Textbooks.
Classification: LCC QL752 .V83 2024 | DDC 591.7/88—dc23/eng/20231213
LC record available at https://lccn.loc.gov/2023049730

A catalog record for this book is available from the British Library.

Special discounts are available for bulk purchases of this book. For more information,
please contact Special Sales at specialsales@jh.edu.

CONTENTS

PREFACE

The goal of *The Biology and Conservation of Animal Populations* is not to broadcast as much material per page as possible. Rather the goal is to teach as many of the important lessons of population biology as can be effectively taught, given the book's length. The decisions I made about what constitutes effective teaching in this textbook were based on three decades of teaching undergraduates and two decades of teaching this subject.

Over that time, my teaching has changed and students' learning has changed. Some changes have emerged slowly; others were catapulted by the COVID-19 pandemic. This book was written conscious of these changes. To help instructors and students use this book for the greatest gain, I'll explicitly share some of the important features and orientations of *Animal Populations*.

MATH

Population biology is fundamentally and inescapably a mathematical subject. The concern is that many students interested in ecology and conservation dislike or are intimidated by math. Those feelings rise, in part, from shortcomings in the teaching-learning process that occur years before students come to study population biology. While that history cannot be rewritten, we can recoup all that may have been lost along the way. I am proof for that claim, inasmuch as I suffered from severe math phobia until the last year of my undergraduate education.

A key step in this recuperation is for instructors to accurately identify where students are in terms of their understanding of math. If you are new to *teaching* population biology, then you might be surprised to know that many upper-class undergraduates do not know what logarithms are or why they are useful, some do not know that a subscript does not represent a mathematical operation, and many cannot solve algebraic equations or interpret graphs with the proficiency that you might suppose.

Many colleges and universities have responded to challenges in math education by offering and requiring remedial math courses for incoming students. Those courses risk making instructors of population biology think, "Problem solved." Unfortunately, there is compelling evidence that it is not. A likely reason is that for many people learning math for its own sake is difficult and unrewarding. More often than not, becoming mathematically proficient requires seeing how math is necessary to solve problems that impress one as being of natural interest or import.

Yet, it is not enough to say that population biology is interesting and important, so let's go. Rather, *Animal Populations* takes the knowledge that students have—or do not have—and builds from there.

For example, portions of this book have the feel of a manual that might accompany a computer lab. The best way for students to understand an equation—be it the equation for exponential growth (Chapter 2) or for projecting the abundance of a structured population (Chapter 5)—is for students to use the equations directly and immediately. I have elected to use Microsoft Excel as the platform for this aspect of instruction because it is widely available to those with access to a computer. But the focus on Excel is easily shifted to any spreadsheet software.

Animal Populations makes an explicit connection between learning the principles of population biology and gaining proficiency in Excel. Doing so is valuable, in part, for providing students with immediate feedback on their learning. If a student can reproduce a result given in the text, then they can gain some confidence about their understanding. This approach also encourages curiosity-based exploration of the behavior of the equations and the populations they represent.

FLIPPED CLASSROOMS AND REMOTE LEARNING

Flipped classrooms are increasingly common, but still not the norm. They involve using time between class meetings for students to receive information and time during class meetings to work with that information. As such, they are an antidote to the limitations of classrooms dominated by lectures and lecture-like broadcasts of facts.

Flipped classrooms aim to capitalize on the distinctive learning opportunities that arise from students engaging with each other in the presence of an instructor. They also capitalize on the increasing availability of high-quality ways that students can first receive information outside of the classroom. One of those high-quality vehicles of information is this book.

It was designed for students' first exposure to the ideas it conveys. The book's value does not depend on its being used in a flipped-classroom context, but it is well suited to doing so.

It is also compatible with remote learning. I wrote *Animal Populations* throughout various phases of the COVID-19 pandemic and used draft chapters to teach face-to-face and remotely. The book works well for both modalities of instruction.

ONLINE SUPPLEMENT

Animal Populations is supported by an online supplement. It includes basic features, such as annotated lists of further readings for each chapter, as well as materials that can form the basis for activities that might occur during class meetings after students have read from the book on their own. A partial guide to the online supplement and how I use its components in a flipped classroom follows.

Chapter Summaries

The online supplement includes a bulleted list of summary points for each chapter. I separated the summaries from the book itself to make them a little less accessible for the purpose of encouraging students to develop their own summaries that can be compared to what is in the online supplement.

Many students are still developing their reading skills, in particular their ability to organize and summarize the large amount of information that appears in a chapter. It is easy for instructors to overestimate students' abilities in these regards.

To gauge and assist students, I routinely use one class meeting for students to work in small groups to develop their own list of the most important summary statements of a chapter. Toward the end of the class meeting, we reconvene as one large group, where the smaller groups share with the larger group the fruits of their labor. Then students can compare their lists to the lists that appear in the online supplement.

This exercise is valuable, not only for students making the knowledge their own, but also for instructors to identify what ideas present the greatest challenge to students.

Discussion Boards

Acquiring new knowledge depends on the ability not only to summarize what one has read, but also to discuss that new knowledge with others. Students need practice discussing new ideas, and instructors can provide space for that discussion during class meetings.

Many institutions have learning management systems (LMS), such as Canvas or Blackboard. I use our institution's LMS to administer discussion boards, prompted by questions that require students to synthesize and process knowledge presented in the chapters. I instruct students to use the discussion board as a primer for small-group discussion that we'll hold during an upcoming class meeting. This ensures that every student has something to share during the class meeting. At a minimum, all a student needs to do is share what they wrote on the discussion board. The process may seem repetitious to an instructor. But it's not. Expressing one's self in writing and then orally are complementary ways of gaining ownership of new knowledge. The online supplement includes the text that I use to prompt many of these discussions.

Practice Problems

Becoming proficient at solving math problems is key to learning population biology. My approach to helping students on this path includes providing sets of *practice problems*, which are distinct from *assessment problems*. For the practice problems, I provide an answer key in a separate file to which they have access. Students are warned not to look at the answer key until they have tried solving the problems. Students benefit greatly from seeing whether they are finding the right solution at a time that matters most to them—namely, when they are working on those problems, as opposed to several days later, after an instructor grades their work. Students submit their solutions for credit, and I assign credit on the basis of their having carefully and thoughtfully presented a solution. (No need to see if they got the right answer, as that information is already available to the students.)

I also provide substantive time during class meetings for students to work in small groups to collaboratively discover solutions to these problems. Students must submit their own work, but they are encouraged to collaborate in the discovery process. In my experience, observing students work on problems during class meetings has been a valuable way for me to discover and mitigate the obstacles they find along the way.

For the assessment problems, no answer key is provided until after I grade the work that students submit. I assign credit for assessment problems primarily on the basis of whether students get the correct answer. After practice problems and assessment problems, students' proficiency is evaluated for a third time, during exams.

The vast majority of students get the idea behind this approach. Students who try to game the system by not taking the practice problems seriously are easily identified by their poor performance on the assessment problems and exams.

The online supplement includes sets of practice problems and answer keys for those problems.

STEM AND INTEGRATIVE LEARNING

STEM is the acronym for science, technology, engineering, and math. Many leaders in education have been singing the virtues of STEM education, and their voices have grown to a crescendo over the past decade. This book—with its focus on the science and mathematics of populations—harmonizes with those voices.

No less important, but somehow less often emphasized, are the virtues of integrative learning through which students connect their newly gained STEM knowledge to other kinds of knowledge. *Animal Populations* pays substantial attention to integrative learning in various ways.

For example, consider the issue of conserving populations in terms of these broad circumstances:

- endangered populations, with numbers that are too low and that some wish to increase
- overabundant populations, with numbers that are too high and that some wish to reduce
- exploited (hunted) populations, whose exploitation serves some worthwhile human interest and for which there is a desire to ensure that the exploitation is sustainable, in the sense of not causing an undue decline in abundance.

Addressing any such case requires *scientific knowledge* pertaining to how and why populations fluctuate in abundance and what kinds of human action alter population abundance. But science is not the only requisite for addressing these cases. Conservation also requires *judgments* that go well beyond science. The best judgments are guided by carefully reasoned values.

Take, for instance population endangerment, which is the focus of Chapter 6. The science of endangerment is all about providing estimates for the extinction risk experienced by a population. That science requires making a critical judgment: Namely, how much risk does a species (or population) need to experience before it deserves special protections? The answer cannot be zero risk, because we'll see that the science indicates that all populations always experience some risk of extinction. So, the question remains: How much extinction risk is too much? Answering the question is greatly aided by knowing the relevant science, but the answering goes far beyond science. Chapter 6 attends the science of endangerment and the value judgments involved.

Exploitation is also an intricate mix of science and values. Its scientific elements include an ability to anticipate how different kinds of populations are affected by different intensities of exploitation. The values at stake are a complex confluence of many considerations, including the cost of exploitation, meaning the lives of the killed animals; the risk of overexploitation, which is a common cause of species endangerment; and proper acknowledgment that exploitation is a purpose-driven activity. The purposes of exploitation vary widely from case to case. Often, but not always, the purpose of exploitation involves supporting vital aspects of some humans' well-being. Consequently, judging the desirability of any case of exploitation requires evaluating its purpose and the likelihood that the planned harvest strategy will realize that purpose. Chapter 8 attends those value-laden facets of exploitation.

Identifying and beginning to attend the connections between science and value judgments are key aspects of the integrative learning that *Animal Populations* teaches. It thus includes an aspect of integration that many leaders in education have called for: the integration of science and ethics.

That lofty ambition needs to be counterbalanced with a harsh reality. It is not possible to adequately cover both the science and ethics of populations in a book this length. Therefore, *Animal Populations* aims to teach the science thoroughly—because that aligns with the kind of classes for which this book is intended—and to build substantive bridges from the science to the values involved in conserving populations. These bridges show students how population biology and conservation are connected to other elements of their curriculum. Where the bridges stop short in *Animal Populations*, guidance for completing those bridges is provided in the online supplement to this textbook.

To summarize, *Animal Populations* fosters the synthesis of (scientific) facts and (ethical) values through formal methods of critical thinking. In doing so, it treats some underattended challenges in higher education and is distinctive among textbooks on population biology.

ORGANIZATION OF CHAPTERS

The motivations for our actions and interests merit more reflection than they often receive. Thus, Chapter 1 is entitled "Why Study Animal Populations?" and offers a pair of reasons for studying populations that definitely merit reflection.

A core motivation of Chapter 2 is the simplest and most far-reaching principle of population biology. Namely, fluctuations in population abundance are best described as percentage changes in abundance. Of course, one can usefully describe a population in terms of absolute changes in abundance, but the best descriptions involve percent changes. In other words, and for example, it's fine to say that this population of 100 individuals grew by 25, but it's far more insightful to say that it grew by 25%. You'll be impressed in Chapter 2 by how much insight comes from that simple principle. If the mathematics of Chapter 2 do not strike you as interesting in their own right, your interest will likely be held by its description of a controversial case of killing elephants to control their abundance.

The second most important principle in population biology is deceptively simple: Populations cannot grow to infinity, and they tend not to go extinct (except when humans drive them to extinction). Populations tend to persist for long periods of time, fluctuating in the wide space between extinction and infinity. In Chapter 3, we'll discover all the important and less-than-obvious implications of that simple idea. This chapter is replete with examples featuring warblers, ibex, marmots, and more.

The examples covered in Chapters 2 and 3 raise plenty of implicit and explicit questions about how we should treat animal populations. Chapter 4 focuses on methods for synthesizing scientific facts with basic ethical claims to better understand judgments about the treatment of populations. These judgments reveal for students the connections between conservation and social justice for humans and nonhuman animals.

Animal Populations does not tell students how to treat populations. Rather it provides access to the formal techniques of critical thinking that are essential for making and evaluating such judgments. The lessons of Chapter 4 are carried throughout the remaining chapters of *Animal Populations*.

Chapter 5 returns our attention to science by inspecting the complicated consequences of another seemingly simple reality: namely, that individuals within a population are not identical. They may differ, for example, in terms of age or social status, and such differences influence a population's dynamics. Thus, two deer populations could have the same abundance, yet still be very different. For instance, one population may be comprised mainly of prime-aged individuals with high rates of survival and reproduction, while the other population may be comprised mainly by senescent individuals with low rates of survival and reproduction. Chapter 5 explores the dynamics of populations that are structured according to their age, reproductive behavior, or social status.

Chapter 6 of *Animal Populations* draws on all the prior chapters to examine the dynamics of populations that are threatened with extinction. Those ideas carry into Chapter 7, which focuses on the influence that genetic diversity (and its loss) have on population dynamics. Both chapters address the question of how large does a population need to be to avoid an undue risk of extinction? Chapter 8 is about the exploitation of animal populations.

Finally, most animals eat other organisms, are eaten by other organisms, or both. The population dynamics of such species are linked in the sense that the source of mortality for the one population is a source of food for the other. Chapter 9 studies the dynamics of populations linked in this way.

ACKNOWLEDGMENTS

So much of my understanding of biology and conservation of populations is due to enduring relationships with friends from whom I've learned so much, especially Doug Smith, Rolf Peterson, Jeremy Bruskotter, and Michael Paul Nelson.

I'm also grateful to Jared Wolfe for his review of Chapter 6, Astrid Vik Stronen for her review of Chapter 7, and Chris Webster for his review of Chapter 8. To David W. Macdonald I am immensely grateful for his review of the entire book. I have unending gratitude for Leah, my wife, who also reviewed and discussed every paragraph of this book.

Andrew J. Storer, Dean of Michigan Technological University's College of Forest Resources and Environmental Science, has been valuable for shaping a professional environment where it is possible to pursue projects such as this.

I am especially grateful to the staff of Johns Hopkins University Press who made my experience with this book a dream. Ezra Rodriguez made the incorporation of illustrations a breeze. I owe much to Tiffany Gasbarrini for believing in this project from the beginning.

Several passages of text in Chapter 9 are excerpted and edited from Vucetich (2021), particularly passages pertaining to Vito Volterra and his equations, the microcosm experiments, and the otter trophic cascades. Highlighting pygmy hippos to illustrate inbreeding depression in Chapter 7 was inspired by Frankham et al. (2002). The idea that genetic drift can be explained with a metaphor involving marbles (Figure 7.5 and associated text) is from the *Understanding Evolution* webpage (https://evolution.berkeley.edu/sampling-error-and-evolution/) by the University of California Museum of Paleontology. All links were accessed on November 10, 2023.

ONLINE SUPPLEMENT

The Biology and Conservation of Animal Populations is supported by an online supplement. It provides summary points, discussion questions, and annotated lists of further readings for each chapter. It also features materials that can form the basis for activities during class meetings, or that students may use on their own. These include Excel demonstrations, practice problems and solutions, and deeper explanations of selected topics. You can find these supplemental materials by visiting press.jhu.edu/ books, searching on the book title or author, viewing the book's page, and scrolling down to the Additional Resources section.

CHAPTER 1

Why Study Animal Populations?

Before starting this chapter, take a moment to read the Preface for information that will help you get the most out of *Animal Populations*.

Animal Populations aims to engage you in learning what is needed to mitigate the biodiversity crisis afflicting the Earth. This crisis is routinely characterized by the rate at which humans are driving whole species to extinction, which is thought to be approximately three orders of magnitude greater than the background rate, or the rate of extinction that would occur if humans were not causing extinctions (Pimm et al. 2014). Because of that elevated rate, an estimated 20% of the 40,000 species of mammals, birds, fish, and reptiles are at significant risk of global extinction over the next century (Hoffman et al. 2010).

Those statistics are grim, but they also represent only one aspect of the biodiversity crisis. The crisis has a second insidious facet—namely, that the global extinction of a species is preceded by many local extinctions of populations as a species' geographic range contracts (Fig. 1.1). The cumulative effect of those local extinctions is that the majority of terrestrial vertebrates have been extirpated from 60% or more of their geographic ranges (Ceballos and Ehrlich 2002; Ceballos, Ehrlich, and Dirzo 2017). That statistic is plenty shocking, but these local extinctions have another staggering and largely overlooked consequence: the breathtaking loss of native biodiversity from most regions of the planet. For example, most native mammalian fauna have been extirpated from large portions of North America and Africa (Fig. 1.1).

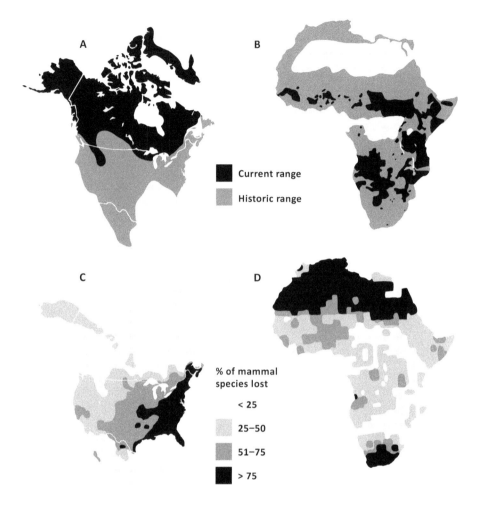

Figure 1.1 The geographic ranges of most species are contracting, as exemplified by wolves (A) and lions (B). These range contractions result from numerous populations going locally extinct. The cumulative effect of so many species experiencing range contraction is that large portions of the planet have lost many of their native species, as exemplified by panels C and D. Reprinted from Vucetich (2021) and adapted from Ceballos and Ehrlich (2002), Bauer et al. (2016), and Bruskotter et al. (2014).

Species have ecological value, which is manifest through the roles they play in maintaining the health and function of native ecosystems. Animal populations cannot fulfill those roles in places where they have been extirpated. Hence, the widespread loss of populations is truly a crisis.

In sum, the biodiversity crisis has two facets, global extinction and the loss of populations. Knowing how to mitigate those losses requires understanding the science and conservation of animal populations.

WHAT IS A POPULATION?

The preceding thoughts may be more than enough to inspire many of you to roll your sleeves up and work hard to study population biology. But there is a more powerful motivation—if you're open to it. This motivation arises from a rich understanding of what the essence of a population is. To share that, I need to tell you a little about myself.

It's been my profession for the past 30 years to make scientific discoveries about populations. Some of these discoveries—the details of which I'll share later in this book—pertain to the wolves and moose that live in Isle Royale National Park, which is a remote island in the northwest corner of Lake Superior in North America.

In pursuit of those discoveries, I have spent seven weeks of most winters during my adult life on Isle Royale estimating the abundance of the populations of wolves and moose. To help explain the year-to-year fluctuations that I've observed, I also estimated the frequency at which the wolves kill moose. This so-called kill rate represents the rate at which wolves acquire food, which wolves need to survive and reproduce. This kill rate also leads to estimates for the predation rate, which is the annual percentage of the moose population that dies from wolf predation.[1] The observations required for these estimates are made from a small aircraft that can hold only a pilot and one passenger. Here is one such observation:[2]

February 18, 2012. Last night the wolves of Chippewa Harbor Pack crossed the Greenstone Ridge to hunt in remote corners of their territory. By morning they were traveling back toward the core of their territory.

Shortly after we caught up with them, not far from Little Todd Harbor, we watched them change course, abruptly and straight into the wind. We saw what they smelled—a cow moose and her calf, who had themselves been foraging. It didn't look good for the cow and calf right from the beginning. The calf was too far away from her mother, and they may have had different ideas about how to handle the situation. The wolves rushed in.

The cow turned to face the wolves, expertly positioned between the wolves and her calf, but only for a second. The calf bolted. After a flash of confusion, the cow pivoted and did the same. Had she not, the wolves would have rushed past the cow and bloodied the snow with her calf. The break in coordination between cow and calf put four or five wind-thrown trees between the cow and her tender love. The cow hurled herself over the partially fallen trunks. She caught up with her frantic calf before the wolves did.

Then the chase was on, led by the least experienced of them all—the calf. The cow, while capable of running faster, stayed immediately behind the calf, no matter what direction the terror-ridden mind of that calf decided to take. Every third or fourth step the cow snapped one of her rear hooves back toward the teeth of death. One solid knock to the head would rattle the life from even a hound of hell.

After a couple of minutes and perhaps a third of a mile, the pace slowed. By the third minute everyone was walking. The calf, the cow, and the wolves. The stakes were high for all, but not greater than the exhaustion they shared. Eventually they all stopped. Not a hair's width separated the cow and calf, and the wolves were just 20 feet away. The cow faced the wolves. A few minutes later the wolves walked away. By nightfall the Chippewa Harbor Pack had pushed on another six miles or so, passing who-knows-how-many-more moose. Their stomachs remained empty.

In a typical year, most moose on Isle Royale are tested by wolves about once a month. No moose fails more than once, and most moose eventually fail the test. They die from circulatory shock, hypovolemia (loss of blood), insufficient blood pressure in the brain, or injury to some other vital organ—all preceded by an ocean of exhaustion, a lifetime of anxiety compressed into a moment held by a fast and shallow heartbeat, a light and vertiginous feeling in the head brought by low blood pressure, and a dissociating numbness that is a true gift, offered by a cocktail of endogenous biochemicals and suggestive that evolution may be more compassionate— if only by accident—than her reputation. Teeth, swinging hooves, bloody snow, spinning sky, faintness, a once proud and still massive shoulder hitting the ground hard, tearing of flesh, fading, and then nothing.

When wolves eat enough moose, wolf births tend to outpace deaths, and the wolf population grows. Otherwise, the population declines. The tipping point occurs when a pack kills and eats about three-quarters of a moose each month for each wolf living in the pack.[3] When wolves kill moose too frequently, moose births cannot outnumber the deaths in a moose population, and the moose population declines. Otherwise, the population grows. The breaking point for moose occurs when wolves kill about 10% or 12% of the moose population in a year. Those are not strict thresholds so much as tendencies that are mitigated by disease, weather, the genetic health of the wolves, and the nutritional condition of the moose.

Every population fluctuates over time. It rises and falls as certainly as organisms respire. Describing and explaining these fluctuations are central concerns of *Animal Populations*. Sometimes the fluctuations are slow and deep; at other times they are shallow and erratic. Every population explosion, every decades-long increase, and every erratic, upward spike is the result of births outnumbering deaths. Every kind of population decline happens when deaths eclipse births.

Fluctuations emerge from an immortal stream of births and deaths. Our humanity requires—while thinking about populations—that we never lose sight that every dewy birth, every bloodied corpse, and every life in between is at once joyous, tragic, and marvelous. While a population emerges from that continuous sequence of births and deaths, populations are also more than the sum of those life-altering events in the same way that the life of an organism is more than the sum of heart beats, breaths, metabolism, catabolism, and all the other discrete functions comprising a living being. When a thing is more than the sum of its parts, scientists say that it has *emergent properties*. These emergent properties are in many ways the subject of *Animal Populations*.

In the opening paragraphs of this short chapter, I have shared a practical reason to study population biology. Now I am sharing what is likely a more powerful motivation, one that follows from the wonder and awe that populations engender among those who are open to these feelings: good science suggests that wonder and awe make one a better student and a better scientist—not to mention a more peaceful, healthy person (Allen 2018; Green 2018).

There will be times during your study of population biology when it all seems to be inert numbers and sterile equations. At those times you need to remind yourself of what a population really is.

Here is another aspect of life on Earth to recognize and appreciate: it is that life on Earth is organized hierarchically. Thus, we have organisms, which are the most recognizable and familiar packages of life. Organisms of the same species living in the same area are populations, and populations of different species living in the same area are communities. Organisms, populations, and communities—those are three of the most basic manifestations of life on Earth. This book is about that mid-tiered manifestation of life: populations.

Communities and their dynamics emerge from interactions among populations. In some cases, populations merely coexist, coincidentally and without much interdependency. In other cases, populations are tightly linked, such as the wolves and moose on Isle Royale. The strength of interactions among populations runs the gamut from loose to intimate. The point is that anyone who wants to understand community ecology will need to understand population biology.

Populations are also deeply affected by the individual organisms of which they are comprised. An individual organism is usefully conceived as a bundle of phenotypes—anatomical phenotypes; life history phenotypes, such as age-specific rates of survival and reproduction; and behavioral phenotypes, including foraging behaviors, anti-predation behaviors, and competitive and cooperative behaviors. Phenotypes are influenced, and sometimes determined, by an organism's genotype. So, anyone who wants to understand populations will need to understand the genotypic and phenotypic characteristics of organisms.

The grand point is simple, albeit easy to underappreciate: populations are nestled between individuals and communities. Populations connect life from one of its most recognizable forms (organisms) to one of its grandest forms (communities).

These understandings of the essential nature of populations may well inspire your curiosity about how populations work and what is required of us to coexist with populations of life on Earth.

To answer the question that forms the title of this chapter, Why might one decide to work hard to study animal populations? Because they are beautiful.

CHAPTER 2

Proportional Growth and Density-Independent Dynamics

Consider African bush elephants. They can change the composition of entire vegetative communities by eating some plants and avoiding others. They perform what are reasonably described as funeral rites for the death of a family member. And they can hear through the ground with their feet. In addition to their ecological, behavioral, and physiological magnificence, elephants can also help us understand population biology, including the key ideas of this chapter.

Another reason to attend elephants is that, on the whole, they are not doing well. Most populations have disappeared entirely, many remaining populations are dwindling, and more losses are sure to follow. Every decade of the past century has ended with fewer elephants. Twenty-seven million elephants are thought to have inhabited Africa as recently as just eight elephant generations ago. What remains is a sprinkling of populations (Fig. 2.1).

While African bush elephants are declining overall, some local populations have done exceptionally well. These populations tend to be in areas where they are better protected from habitat destruction and poaching. One such population lives in Kruger National Park of South Africa, where elephant abundance increased to the point that park managers became concerned that elephants would threaten the health of the park's vegetation. That concern turned to action, and thousands of elephants were culled in an attempt to reduce their abundance. The decision was, not surprisingly, controversial. The rationale for culling depended on a certain

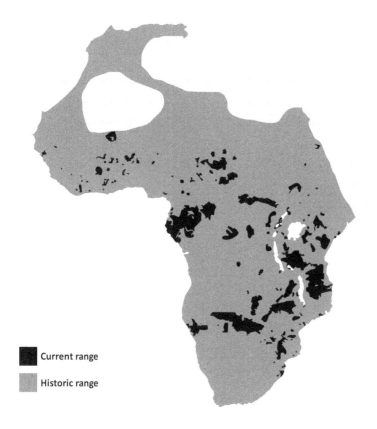

Figure 2.1 The current and historic range of African bush and forest elephants (*Loxodonta africana* and *L. cyclotis*). Adapted from https://commons.wikimedia.org /wiki/File:African_Elephant_distribution_map.svg. The adaptation was to remove certain details from the original, such as national boundaries and bodies of water. Adaptations were made by Jessie Stapleton.

understanding of population biology blended with a particular weighing of human values. Criticisms of culling arose from different understandings of population biology and different sets of values—namely, concerns for healthy ecosystems and for treating individual elephants with dignity.

Outside of protected areas elephants have been declining due to habitat destruction, which is often linked to converting savanna into agricultural lands. That land conversion is often associated with humans living in close proximity to elephants, who sometimes destroy crops and injure or kill humans.[1] When a population of elephants causes chronic trouble for a community of humans, the elephants are typically said to be overabundant (though it's humans that must seem overabundant through elephant eyes). A common response to such conflicts is to kill elephants.

Not enough elephants and too many elephants—both at the same time. Concern for the decline of elephants led the government of Botswana to ban elephant hunting in 2014. The government's leadership changed, revoked the hunting ban, and allowed for about 300 elephants to be killed in 2021 (BBC 2020). Some conservation-minded people lauded the decision, and other conservation-minded people criticized the decision. Reactions differed according to differences in how people synthesized pertinent values and science.

Endangered populations and overabundant populations: the most basic concerns that permeate conservation are too many and not enough. Too few individuals and a population becomes vulnerable to extinction or unable to perform its ecological function. Too many individuals and a population can become trouble for ecosystem health or create conflicts with basic elements of human well-being.

Responding to these concerns requires sharp knowledge of the science (population biology) and incisive understanding of the values that guide our application of the science. It's impossible to arrive at a judgment about how to treat a population without both science and values. We'll begin talking about the values that pertain to populations in Chapter 4. For now, let's get into the first bits of science.

PROPORTIONAL GROWTH

A central goal of population biology is to understand how and why populations fluctuate in abundance. Doing so requires an ability to:

- describe past fluctuations (or dynamics) with useful accuracy and precision.
- evaluate ecological mechanisms that may have given rise to those dynamics.
- make useful predictions about future dynamics.

In view of developing these abilities, let's consider some details about the elephants of Kruger National Park (NP), which was established in 1926 in response to widespread and excessive hunting of wildlife during the late nineteenth and early twentieth centuries. If you are from the United States or Europe, you'll appreciate the fact that the park is about the size of New Jersey or Slovenia. In Kruger NP's early years, fewer than 50 elephants lived within the park. Then, field surveys conducted in 1945 and 1946 indicate that the park population had grown to about 500 elephants.

Now envision the lives of individual elephants as they may have been in Kruger during the mid-twentieth century. Elephants are long-lived, and surviving their first year of life means a good chance that they will live to 55 to 65 years of age. With that life expectancy, only a small percentage of elephants in a population would be expected to die each year. It turns out that percentage is about 2% for a population that is not limited by food or poaching. In other words, such a population has a mortality rate of 0.02/year.

Longer-lived creatures tend to reproduce at slower rates. Among elephants, most females begin to reproduce when they are about 12 years of age, and they can give birth to just a single offspring about once every four years. Some of those 200-pound bouncing baby elephants will not survive their first year of life. Taking account of those life history patterns, one would expect only a small percentage of the population to give birth to offspring that survive to their first birthday. That percentage is about 7% for an elephant population that is not limited by food or poaching. In other words, such a population has a recruitment rate of 0.07/year. *Recruitment* is the term population biologists use to refer to the combination of two biological processes: birthing or hatching and survival of the new offspring for some initial period of time, say a year.

In the previous two paragraphs, we took simple life history observations about individual elephants and scaled them up to population-level statistics. For now, just note three numbers: 7%, 2%, and 500. We'll assume that the abundance of elephants in Kruger NP in 1946 was 500,[2] its mortality rate was 2% per year, and the rate at which new individuals were recruited into the population was 7% per year.[3]

Continue imagining the details of this population by supposing that every elephant in the population has all the food and water they need. Further, suppose that exogeneous sources of mortality[4]—that is, sources of mortality that originate from outside the population—such as humans or lions are at low levels. If it were otherwise, the mortality rate would be higher, and the recruitment rate would be lower.

We can use those numbers (7%, 2%, and 500) to predict how many elephants there would be one year later, in 1947. The prediction works this way: with 500 elephants, we expect the number of new elephants to be 35 ($= 500 \times 0.07$) and the number of elephants to die to be 10 ($= 500 \times 0.02$).[5] So, we'd expect the number of elephants in 1947 to be 525 ($= 500 + 35 - 10$).

Now let's perform those calculations one more time to predict how many elephants there would be in 1948. The arithmetic is a little tedious, but it's also easy.

Having predicted 525 elephants for 1947, we expect the number of newly born elephants for that year to be 36.8 ($= 525 \times 0.07$) and the number of elephants to die to be 10.5 ($= 525 \times 0.02$). So, we'd expect the number of elephants in 1948 to be 551.3 ($= 525 + 36.8 - 10.5$).

Table 2.1 Mathematical prediction of elephant population dynamics in Kruger National Park, 1946–1957. Abundance had been estimated from field surveys in 1953 and 1957. Those estimates are given in parentheses. Notice how close the mathematical predictions are to the field-based estimates. Recruitments refer to the number of new offspring that are born and survive their first year. Column F, which is also labeled, r, is the proportional change in abundance from the current year to the next year, or equivalently from year t to $t+1$. The first value in column F can be calculated as $r = (525-500)/500 = 0.05$. Density refers to the number of elephants per 100 km².

A	B	C	D	E	F	G
Year (t)	Abundance at beginning of year t	Recruitments	Mortalities	Change in abundance from year t to t+1	r	Density
1946	500	35	10	25	0.05	2.55
1947	525	36.8	10.5	26.3	0.05	2.68
1948	551.3	38.6	11.0	27.6	0.05	2.81
1949	578.8	40.5	11.6	28.9	0.05	2.95
1950	607.8	42.5	12.2	30.4	0.05	3.10
1951	638.1	44.7	12.8	31.9	0.05	3.25
1952	670.0	46.9	13.4	33.5	0.05	3.41
1953	703.6 (702)	49.2	14.1	35.2	0.05	3.59
1954	738.7	51.7	14.8	36.9	0.05	3.76
1955	775.7	54.3	15.5	38.8	0.05	3.95
1956	814.4	57.0	16.3	40.7	0.05	4.15
1957	855.2 (855)	59.9	17.1	42.8	0.05	4.36

Did I just predict 551.3 elephants? Yes. But that's okay. While elephants come only in whole numbers (551 or 552 elephants are possible), it's okay that the math offers a prediction that is, strictly speaking, impossible. The math is not intended to be a perfect mirror of the real world. It's intended to be only good enough to be useful. Nevertheless, you are right to have taken notice that the answer is, strictly speaking, unrealistic. An important and challenging skill in population biology is to recognize when and to what extent a mathematical prediction is unlikely to be *adequately* accurate. For now, if the math predicts 36.8 births and 10.5 deaths, it's okay to expect that there'd be 36 or 37 offspring and 10 or 11 deaths.

Take a look at Table 2.1, specifically columns A through E, to see the predictions I've made for the number of elephants in Kruger for each of 11 years, from 1947 to 1957. I made the predictions in Excel using calculations just like those described above. Scan column B of Table 2.1 to find the two numbers in parentheses. Those are field-based estimates of abundance made during 1953 and 1957. Notice that the predicted abundances are close matches to abundances that were actually estimated in Kruger NP.

> The online supplement guides you in building your own version of Table 2.1. The best time to do so is right after reading this section. Otherwise, Table 2.1 is difficult to understand.

Now take a look a column F. The column's heading is simply r, which is a symbol used by population biologists to represent one of the most important ideas in all of population biology: r is the proportional change in abundance from the current year to the next year. r is also known as the *per capita growth rate*. Table 2.1 demonstrates how r is calculated.

For now, notice that each value of r is 0.05, which means that the population is predicted to have increased by 5% each year. It is no coincidence that this 5% growth rate is equal to 7% (birth rate) minus 2% (mortality rate).

Table 2.1 represents our first effort in developing two of the three abilities presented at the beginning of this section. First, we made a *prediction* about population abundance from 1947 to 1957 based on some assumptions. Second, there is at least some reason to think the predictions are a useful *description* of population dynamics during that time, at least insofar as the predictions for 1953 and 1957 are a close match to the field-based observations for those years. If we accept the claim that Table 2.1 represents a good description of the dynamics, then that description would be:

> The elephant population in Kruger grew at a rate of about 0.05 each year from 1946 to 1957. Over that entire time period, the population grew by about 70% from ~500 to ~855 elephants.

Table 2.1 represents another important idea that applies to all populations, under all circumstances. Namely, population dynamics are described in terms of proportional changes. These proportions can be represented as percentages of total abundance, as when we refer to 2% of the population dying, when we say that the number of newly recruited individuals represents 7% of the population's total abundance, and when we conclude that the population was growing at 5% per year.

These proportions are also *rates*, in the sense that they represent changes over time, such as 2% of the population dying throughout the course of a year. For this reason, it is good practice to say, for example, that the mortality rate is 0.02/year. Keeping track of the units of time is important.

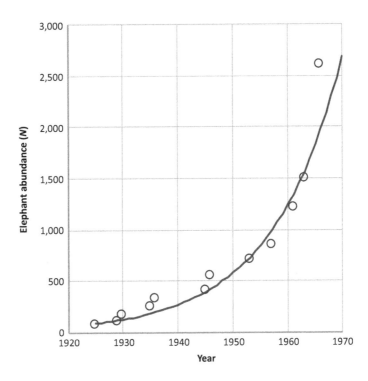

Figure 2.2 Field-based estimates of abundance (circles) for elephants in Kruger National Park, 1925–1966, along with predictions (line) based on assumptions described in the main text. The predictions are close, except for the last year, when the field-based estimate (2,612 elephants) was much higher than the prediction (1,976). That discrepancy could, in principle, be due to an inaccurate field estimate, an important change in the population such that it no longer matches the assumptions used to make the prediction, or both. The field-based estimates are from Whyte, van Aarde, and Pimm (2003).

This idea is important enough to repeat: temporal fluctuations in abundance can be usefully described in terms of proportional changes over time.

The predicted abundances given in Table 2.1 are also depicted in Figure 2.2 along with field-based estimates of abundance from Kruger NP during the same period. The practice problems in the online supplement provide you with an opportunity to reproduce Figure 2.2.

Before moving on, notice the last column of Table 2.1, which reports the predicted density of elephants in Kruger NP. The density of a population is simply the number of animals divided by the area over which the animals are living or the area taken by biologists to be the geographic boundaries of the population. Column G of Table 2.1 is simply column B divided by the size of Kruger NP, which is 19,623 km² (7,576 mi²).[6] The reason for drawing attention to density is that the population dynamics in Table 2.1 and Figure 2.2 are referred to as *density-independent dynamics*.

You've already noticed that the value of r is the same for every year in Table 2.1. That same, constant value of r is also represented in Figure 2.2. No matter what year it was, no matter what the density of elephants had been, the value of r was the same. In other words, r is independent of the population's density. The population dynamics are said to be density-independent.

Populations exhibiting density-independent growth increase slowly at first, but then increase very rapidly, as illustrated by Figure 2.2. When r is a positive number and constant over time, then change in abundance follows a nonlinear, upward-curving trajectory.

ECOLOGICAL PROCESSES

Density-independent dynamics arise from particular ecological processes. Specifically, when every individual in a population has all the resources they need (such as food, water, and shelter) year after year, then birth rate and death rate are determined by physiological limits on life expectancy and reproduction. So long as those conditions hold, per capita growth rate is density-independent and population abundance increases nonlinearly, as illustrated in Table 2.1 and Figure 2.2.

These ecological conditions are often met for a limited time after a population colonizes a new, large area of suitable habitat or as a population recovers from a major catastrophe, as when the formation of Kruger NP allowed the elephant population to recover from excessive hunting.[7] Density-independent dynamics can also be useful for understanding the expansion of non-native or invasive species, as well as endangered species in the process of recovering.

A population may be exposed to exogenous sources of mortality (such as predation, hunting, or disease) and still exhibit density-independent dynamics, so long as the proportion of individuals that die from all causes combined is more or less constant over time or any change in mortality rate is offset by a compensating change in the recruitment rate.

No population can exhibit density-independent growth for very long periods of time. Any population that did would grow toward infinity. So, when one observes density-independent growth, the questions to ask are, how long will that growth continue, what ecological process will intervene to stop it, and what will the abundance be when density-independent growth shifts to some other kind of population dynamic?

While this chapter is focused on constant r, much of this book is an exploration of the ecological processes that cause r to fluctuate over time. The most important causes of fluctuation include intraspecific competition, predation, disease, harvest by humans, and fluctuations in the availability of food and abiotic environmental conditions, especially climatic fluctuations.

SOME EQUATIONS

It is almost impossible to understand the influence of various ecological processes on population dynamics without taking advantage of mathematical equations. We'll begin gently by exploring three equations that represent the population dynamics we've been considering.

Knowing the reason for using equations is a big step toward making that effort a little easier, and the reason is that equations are a great way to explain complex ideas with great precision and succinctness. Think of translating a complex idea from English words into a more compact language.

The precision and succinctness offered by equations comes from replacing English words and phrases with simpler symbols. We've already done an important bit of this translating from English to math when we let the symbol r stand in place for "proportional change in abundance from one year to the next."

Let's use another symbol, N_t, to represent "abundance or density at some time." Hence, the symbol N_{t+1} can refer to "abundance or density at the next time." The letter N refers to abundance and the subscripted letters and numbers indicate the time. Using those symbols, we can write:

$$N_{t+1} = N_t + N_t r_t. \tag{2.1}$$

This equation can be read like an English sentence whose meaning is:

Next year's abundance (N_{t+1}) is equal to this year's abundance (N_t) plus this year's abundance (N_t) multiplied by the proportional change in abundance (r_t).

As with all equations, if you have all the information on the right side of the equation, then you can calculate the value of the equation's left side. For example, if there were 525 elephants (as in 1947) and if the growth rate was 0.05/year, then you can replace the symbols in Equation 2.1 with numbers to calculate the predicted number of elephants in 1948: $N_{1947+1} = 525 + (525 \times 0.05) = 551.3$. From that example, you can see how column B of Table 2.1 can be calculated from column F.

Equation 2.1 can also be algebraically rearranged so that r_t is by itself on the left side of the equation:

$$r_t = (N_{t+1} - N_t)/N_t. \tag{2.2}$$

Take a look at the note that appears with Table 2.1, and you'll see that we've already used Equation 2.2. You'll also get a rich sense for the importance of Equation 2.2 with the practice problems of the online supplement.

Equations 2.1 and 2.2 are definitely worth memorizing. Better yet, your instructor and I will give you enough practice with these equations that you won't need to memorize them, because with practice they will begin to make intuitive sense.

r_t can be expressed in terms of abundance, as shown in Equation 2.2. It can also be expressed in terms of its two constituent parts, the recruitment rate and the mortality rate. It's common to let mortality be represented by the symbol d (think death). And it's common to let recruitment rate be represented by the symbol b (think birth). More specifically, per capita growth rate is equal to the recruitment rate minus the mortality rate (which are columns C and D of Table 2.1):

$$r_t = b - d_t. \tag{2.3a}$$

Equation 2.3a is the reason why it is no coincidence that in Table 2.1 the recruitment rate is 0.07/year, the mortality rate is 0.02/year, and the per capita growth rate is 0.05/yr. That is, $0.05 = 0.07 - 0.02$.

There are real-world cases in which a population's dynamics are affected by animals leaving or arriving into the population by dispersal. In those cases, Equation 2.3a should be replaced with

$$r_t = b_t + i_t - d_t - e_t, \tag{2.3b}$$

where i_t represents the rate at which individuals immigrate into the population, and e_t represents the rate at which individuals emigrate out of the population.

Finally, a word about symbols. Students who see symbols such as N_t for the first time sometimes mistakenly (but quite understandably) think that the N and t are connected by some mathematical operation, such as multiplication. This is not the case. When the t appears as a subscript, the N and t are taken to be a single idea. This use of symbols allows for flexibility in expressing ideas. We can write N to represent "abundance" at some unspecified time, or N_t to represent "abundance at time t" (that is, "abundance at some particular time"). So, the subscript simply specifies which abundance we're talking about. We can write, for example, N_3 to represent abundance for year 3 in some series of years. Or we can write N_{2018} to represent abundance in the year 2018. If our meaning is clear without referring to a particular time, then it is okay to suppress the subscripts and simply write, for example, N or r.

Cranes and Wolves

To appreciate the usefulness of Equations 2.1 and 2.2, we can get help from some whooping cranes and wolves. First, whooping cranes.[8] In North America humans drove whooping cranes to the brink of extinction by overhunting and converting their habitat (wet prairies and fields) to agricultural lands. By 1941, there were only 15 whooping cranes. They lived in a single migratory population that overwintered in Texas near the Gulf of Mexico and bred in northern Canada.

Efforts to improve whooping cranes' chances for survival have been underway for more than half a century. The result of those hard-earned conservation efforts is the impressive increase in the abundance of whooping cranes depicted in panel A of Figure 2.3.

A portion of the data used to make panel A is presented in Table 2.2. What's important about Table 2.2 is that r_t is calculated from N_t and N_{t+1} with the help of Equation 2.2, and r_t is the basis for panel B of Figure 2.3.

Notice that in Table 2.2 the last cell in the column for r_t is missing. This is because you cannot calculate r_t unless you know N_t and N_{t+1}. In other words, because the value of N_{1999} is not provided in Table 2.2, one cannot calculate r_{1998}.

In that column of Table 2.2 for r_t, you may have noticed that no two numbers are the same. And while most of the values are positive, there is even one negative value. Given these fluctuations in r_t, what we'd really like to know is:

- Are there any trends in r_t over time or with abundance?
- If there are no trends, what is the average r_t?

Some answers may be found in panel B of Figure 2.3, which suggests no significant tendency for r_t to change as N_t increases. In Chapter 3, we explore more statistically rigorous methods of assessing trends. For now, it's okay to rely on a casual assessment. Panel B also highlights the average value of r_t.

Panel C of Figure 2.3 shows the number of wolves during the first 25 years of their having recolonized Upper Michigan after a century-long campaign of persecution by humans. The wolf and crane populations represent a useful occasion for comparison and contrast. Both populations are clearly increasing. But the lower panels of Figure 2.3 reveal an important difference. While the whooping cranes exhibit

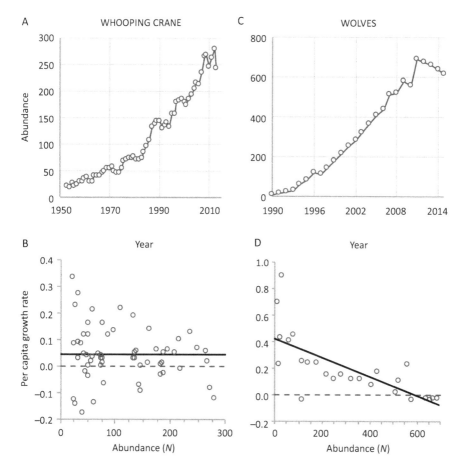

Figure 2.3 Dynamics for a population of whooping cranes (A, B) and wolves (C, D). Panels B and D illustrate density-independent growth and density-dependent growth, respectively. The solid line in B represents the average value of r (which is 0.047/year). The solid line in D is a trend line showing how the expected value of r declines with increases in N. In the lower panels, observations above the dashed line (r = 0) represent years of increasing abundance, and observations below represent years of decreasing abundance. Data for panel C is from the Michigan Department of Natural Resources (2022).

density-independent growth (panel B), the wolf population exhibits density-dependent growth, where r_t tends to decrease as N_t increases (panel D). Chapter 3 is devoted to density-dependent growth. For now, it is enough to know that density dependence stands in contrast to density independence.

Now is a good time to visit the online supplement where you'll find practice problems that involve reproducing Figure 2.3.

Table 2.2 Estimated abundance from a population of whooping cranes, 1991–1998. N_t are field-based estimates of abundance, and r_t is calculated from N_t and N_{t-1} using equation 2.2. For example, the value 0.030 is equal to (136–132)/132, meaning that the population grew by 3.0% from 1991 to 1992.

Year (t)	N_t	r_t
1991	132	0.030
1992	136	0.051
1993	143	−0.070
1994	133	0.188
1995	158	0.013
1996	160	0.138
1997	182	0.005
1998	183	

As mentioned, the solid lines in the lower panels of Figure 2.3 convey the general tendencies of r_t. Is anything to be made of the fluctuations around those general tendencies? For example, the crane population seems to exhibit more fluctuation around its solid line than the wolf population. And r_t seems to have been more variable when the abundance of cranes was low (left side of panel B). For now, I want to point out only that these fluctuations are important and that they receive more attention in Chapter 3.

EXPONENTIAL GROWTH

Equation 2.1 is valuable for being able to represent proportional growth if r is constant or if it fluctuates over time. Another equation, called the *equation for exponential growth*, is especially convenient for the specific case where r is constant over time:

$$N_t = N_o e^{rt}. \tag{2.4}$$

The symbols N_o and N_t represent abundances at two times. Think of N_o as initial abundance and N_t as abundance after some time (t) has passed. If we know N_o and r, then Equation 2.4 allows one to predict N_t with a single calculation, even over long periods of time.[9]

Before putting this equation to use, let me draw your attention to what may seem its weirdest part, the symbol e. It represents a special number, equal to approximately 2.718. You likely learned about e in a math class from high school or first year in college. If you want to recall what you learned, type "e mathematical constant" into an internet search engine or YouTube. In any case, it's good enough for now just to know that e is a special number. And we will be raising the number e to the power of another number, and that other number is the product of time (t) multiplied by the per capita growth rate (r).

Equation 2.4 can be usefully rearranged so that r is by itself on the left side:

$$r = \ln(N_t/N_0)/t, \tag{2.5}$$

where $\ln()$ tells us to take the natural logarithm of the quantity within the parentheses. With that equation you can calculate the growth rate of a population if you know N_0, N_t, and t.

Equations 2.4 and 2.5 can be used together to describe past dynamics and to predict future dynamics. For example, there were an estimated 24 whooping cranes in 1953. Sixty years later (in 2013), there were an estimated 245 cranes. According to Equation 2.5, the value of r during that time was 0.039 per year ($= \ln(245/24)/60$).

If the population keeps growing at that rate over the next 50 years, then there will be 1,698 cranes ($= 245e^{0.038 \times 50}$) in the year 2063. To perform this calculation you can type "$= 245*\text{EXP}(0.038*50)$" into any cell of a Microsoft Excel file.

To put 1,698 cranes into context, it is believed there were 15,000 to 20,000 whooping cranes before humans began reducing their abundance. Equation 2.4 can be rearranged to predict how much time it would take to reach those levels, given values of r and N_0:

$$t = \ln(N_t/N_0)/r. \tag{2.6}$$

For this crane example, the calculation to perform is $\ln(15{,}000/245)/0.038$, which is equal to 106 years.

Calculations like these reveal an important reason why restoring diminished populations takes time and commitment.

Declining Populations of Deep-Sea Fish

The ideas we're discussing apply not only to increasing populations but also to decreasing populations. In that case, one merely inserts negative values of r into equations 2.1 or 2.4. And equations 2.2 and 2.5 will produce negative values of r when they are fed values of N that represent population decline (e.g., see the value for r_{1993} in Table 2.2). Simple as the concept may be, we'll benefit from an example.

After decades of overfishing waters above the continental shelf in the northwest Atlantic, several important fisheries collapsed in the 1960s, 1970s, and 1980s. That collapse drove many fishers further out and into much deeper waters.

The species of fish that are distinctive to deep-sea habitats tend to be long-lived, grow slowly, and reproduce slowly. This life history makes them vulnerable to overharvest—even if killed merely as by-catch in the process of fishers targeting other fish such as halibut.

While the shift to fishing in deeper seas was concerning from the beginning, data to assess the concern are difficult to acquire. In 2006, researchers analyzed survey data from several species of deep-sea fish that had been collected between 1978 and 1994.[10]

One of these species is the onion-eye grenadier (*Macrourus berglax*). This fish feeds on crustaceans, sea worms, and small fish at the ocean floor in ice cold waters, 600 to 6,000 feet beneath the surface. They have large heads with pointy snouts and mouths that protrude from the bottom side of their heads. They seem mis-proportioned in that their heads are about as long as their trunks, and their tails are as long as the rest of their body. The tail begins as thick as their trunk and gradually tapers to a point. Their bodies are also covered with gray scales, many of which sport tough spines. Their stunning eyes are conspicuous enough to give them their name. And, they can grow to 3 feet in length.

Figure 2.4 shows the temporal trend in an index of abundance for these onion-eye grenadiers. Clearly the population had been declining, but these researchers had two more specific goals: first, to quantify the rate of decline; second, to predict how long it would take for the population to recover if they were protected from overfishing.

With respect to the first goal, I offered a simple way to quantify r when I introduced Equation 2.5. But there is a better way to estimate r if one has an entire time series of data. Showing you this better way requires stepping away from deep-sea fish for a few paragraphs.

Log-Transformed Abundance

The idea we need is readily explained by recalling the contrast we observed between the populations of whooping cranes and wolves (Fig 2.3). Both populations increased over time, but only the crane population exhibited exponential growth. Here is a quick and easy way to distinguish exponential changes in abundance from other kinds of change.

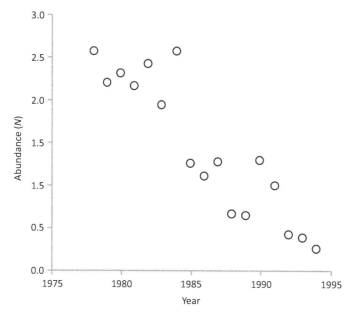

Figure 2.4 Decline in abundance of the onion-eye grenadier. Abundance in this case is an index of abundance, based on the number of fish caught in a survey trawl, scaled to account for how long the trawl was dragged across the ocean floor.

Start with Equation 2.4 and suppose we are interested in predicting abundance just one unit of time (say years) into the future, so that $t = 1$. Because 1 multiplied by anything is itself, we can rewrite Equation 2.4 without the t, like this:

$$N_{t+1} = N_t e^r.$$

Now take the natural log of both sides of the equation:

$$\ln(N_{t+1}) = \ln(N_t e^r).$$

By the rules of logarithms that you might recall from an earlier math class, the right side of the equation can be rewritten as

$$\ln(N_{t+1}) = \ln(N_t) + r. \qquad (2.7)$$

Equation 2.7 says that $\ln(N)$ increases by a fixed amount with the passage of every fixed unit of time, and that amount of increase is r. In other words, a population whose abundance (N) is growing exponentially will also exhibit a linear rate of growth in $\ln(N)$. If you enjoy the mathematical aspects of population biology, then that idea is worth careful reflection. If not, that's okay, because what's more important is that you understand what is depicted in Figure 2.5.

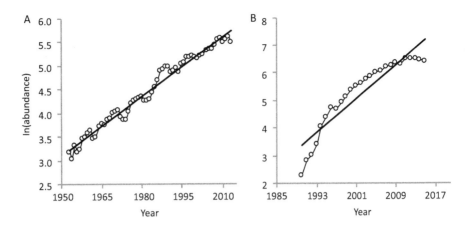

Figure 2.5 Log-transformed abundance for the same populations of whooping cranes and wolves, whose untransformed abundances are depicted in Figure 2.3. If a population (like that of the whooping cranes) is growing exponentially, then a time series of log-transformed abundance will increase linearly (A). If a population's growth rate is slowing down over time, as is that of the wolves, then the log-transformed abundance has a concave shape (B). If the data are well-represented by a linear trend (as in panel A), then the slope of the trend line is an estimate of r. The trend line in panel A has the formula $ln(N) = -79.25 + 0.042(year)$. As such, the estimate of r is 0.042 per year (or 4.2% per year).

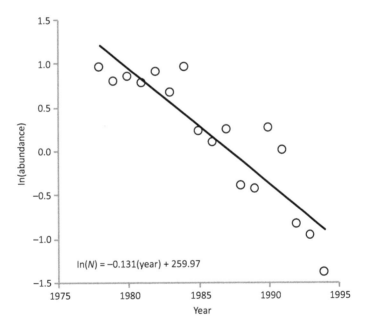

Figure 2.6 Log-transformation of the data presented in Figure 2.4. Because this trend is linear, the population is reasonably described as declining exponentially. The equation for the trend line is reported, and the slope of that line is an estimate of the rate of decline.

Back to the Deep-Sea Fish

Those ideas about log-transformed abundance in Figure 2.5 apply to declining populations as well. If log-transformed abundance declines linearly over time, then the population is declining exponentially and Equation 2.4 is applicable.

To quantify the rate of decline (r) for the onion-eye grenadier population, the explanation in the legend to Figure 2.6 is key.

> As you may be coming to expect, the online supplement guides you through this kind of calculation.

If one takes the abundance data for the grenadier population and log-transforms it and then determines the equation for a trend line of that log-transformed data, then the estimate of r is -0.131 per year (Fig. 2.6).

For context, that rate of decline is severe enough to place this population on the list of endangered species that is maintained by the International Union for Conservation of Nature (IUCN). A population declining at that rate is reduced by half every five years.

Recall how toward the beginning of the chapter we used Equation 2.3a (which is $r = b - d$) and life-history information about elephants to infer their per capita growth rate. The researchers who analyzed this deep-sea fish data used an approach that is similar (in spirit, albeit more technical in nature) to estimate the value of r that these species would exhibit if they were no longer overfished. With such an estimate of r and Equation 2.6, one can calculate how long it would take for the population to recover from the past several decades of being overfished.

Because so little is known about the life history of these fish, the best assessment of recovery time is very imprecise. It could be as quick as 20 years and—quite plausibly—as long as 250 years.

The researchers used these findings and similar findings from other species of deep-sea fish to make a compelling case that now is the time to do a much better job protecting the deep sea from overfishing.

CHAPTER 3

Density-Dependent Dynamics

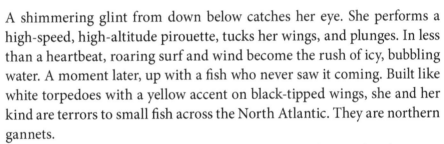

A shimmering glint from down below catches her eye. She performs a high-speed, high-altitude pirouette, tucks her wings, and plunges. In less than a heartbeat, roaring surf and wind become the rush of icy, bubbling water. A moment later, up with a fish who never saw it coming. Built like white torpedoes with a yellow accent on black-tipped wings, she and her kind are terrors to small fish across the North Atlantic. They are northern gannets.

These large diving birds breed in colonies across the North Atlantic on remote, rocky coastlines. Colonies include anywhere from a few hundred to a few thousand birds. After eggs hatch, parents' days are filled with foraging runs out to sea to catch fish for the gaping mouths of their squawking chicks.

When the colony is small, each parent makes short runs for fish—not far out and back shortly. As the colony grows, at least some parents resort to longer runs, which cost more energy and consume more time (Fig. 3.2, panel A). Eventually the colony gets large enough, and no bird has enough time or energy to travel any farther for fish. As that happens, fewer chicks are fledged, and the colony's abundance eventually declines.

This is an example of density-dependent foraging behavior—density-dependent because the behavior changes as the colony's size (or density) increases. Population biology is always connected to the ecological lives of individual organisms. As the colony grows, there may be less food per individual, or it may take more effort to get food, resulting in reduced birth

Figure 3.1 A soaring northern gannet. Credit: Andreas Trepte, www.avi-fauna.info. Source: https://commons.wikimedia.org/wiki/File:Northern-Gannet-8.jpg.

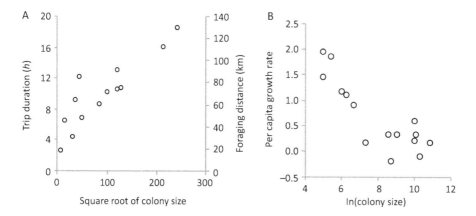

Figure 3.2 Density-dependent behavior and population dynamics for northern gannets. As the size of a northern gannet colony increases, the duration and distance of foraging trips out into the ocean tend to increase (A). Those increases are concomitant with a decrease in per capita growth rate (B). Each data point in these graphs refers to birds from different-sized gannet colonies from the coast of Great Britain. Adapted from Lewis et al. (2001).

rates or increased death rates. Either way, the per capita growth rate will tend to decline with increasing colony size (Fig. 3.2, panel B). This mechanism is called *negative density dependence*, and it prevents populations from growing to infinity.

In Chapter 2 we examined cases in which per capita growth rate (r) was constant over time and did not vary with changes in abundance (N). Now, we'll explore another important case, where r tends to decline as N increases.

DENSITY-DEPENDENT POPULATION GROWTH

Given our interest in making precise descriptions of past dynamics and useful predictions about future dynamics, we'd like mathematical equations that can describe density-dependent population dynamics. Rather than merely present the equations for you to memorize, I'll show you how to build them from easy-to-understand principles. The first step requires analyzing some population data. But before any analysis, data have to be collected.

As an example, we'll take black-throated blue warblers, which spend their summers in the temperate forests of eastern North America. Gathering population data on these warblers is straightforward, at least in principle, because the males are so audibly conspicuous as they incessantly belt out their buzzy, seven-note song. We call it a "song," though "battle cry" might be more apt. The song is a stern warning to other males that this little patch of forest is already taken.

Spend some time in the forest, hear a song, spot the singer, and follow the bird until he takes you to his nest. Set up a mist net nearby, wait for the bird of interest to get caught in the net, attach bands with a unique combination of colors to the bird's legs, and now you can keep track of that particular bird. By doing so with several birds, you can map their (largely nonoverlapping) territories. If you do this for a tract of forest, you can slowly but surely grow a time series of density estimates (number of adult birds divided by the forest area that you are monitoring). Researchers have been doing just that for decades in a forest of New Hampshire (Sillett, Rodenhouse, and Holmes 2004). A fruit of their labor is shown in Figure 3.3.

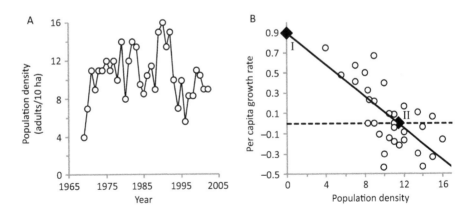

Figure 3.3 Temporal fluctuations in density for a population of black-throated blue warblers from a forest in New Hampshire (A). Those fluctuations correspond to a tendency for per capita growth rate to decline with increasing density (B). See text for explanation of the labeled diamonds. Adapted from Rodenhouse et al. (2003).

Notice how the population's density has fluctuated up and down over the years without any discernable pattern other than that the population shows no risk of growing to infinity or declining to extinction (Fig. 3.3, panel A). But the population *is* following a distinct pattern, and this pattern appears if we look at the same data differently. Make a graph with population density (N) on the x axis and per capita growth rate (r) on the y axis, and see how r tends to decline with increasing N (Fig. 3.3, panel B).

Take a moment to appreciate some of the finer points of panel B. Every data point below the horizontal dotted line represents a year when abundance declined ($r < 0$), and every data point above the dotted line represents a year when abundance increased ($r > 0$). The solid, trend line also features two special points:

· Point I, where the trend line meets the y axis, is the highest per capita growth rate that one can expect from this population. We'll use the symbol r_{max} to represent this rate. Notice that r_{max} occurs when abundance (N) is near zero. In other words, point I is located at coordinate ($0, r_{max}$). r_{max} is sometimes usefully thought of as the per capita growth rate that occurs when every individual in a population has all the resources they need or when there is no intraspecific competition.
· Point II, where the trend line crosses the horizontal dotted line, is the value of N for which one expects r to be zero. For this abundance, subsequent changes in abundance are expected to be zero. We'll designate this special abundance with the mathematical symbol K. The coordinate for this point is ($K, 0$).

To have identified two interesting points on this line is important because—as you might recall from an earlier math class—from any two points on a line one can build the equation that describes the entire line. We're motivated to find this equation because it will be a precise description of the population's past dynamics and basis for predicting future dynamics.

To begin building an equation for r_t, start with the equation that describes any straight line:

$$y = b + mx,$$

where b is the y intercept, m is the slope, and x and y are coordinates on a graph. Replace x and y in that equation with symbols for the ideas represented by the x and y axes of panel B of Figure 3.3. That is, replace x with N_t and replace y with r_t. The result is

$$r_t = b + mN_t.$$

Then, replace b with what we know to be the y intercept of panel B. Specifically, replace b with r_{max}, like this:

$$r_t = r_{max} + mN_t.$$

The next replacement takes a few steps. We need to represent the slope of the line (m) with symbols corresponding to more ecologically oriented ideas. Here's how. You may recall that the slope of a line is also represented by the idea of *rise-over-run*. That is, how much does the line rise (or fall) along the y axis as one moves from left to right along the x axis. Inspect panel B to confirm that as one "runs" along the axis for N, from zero to K, the line falls from r_{max} to zero. That is, the slope is $(0 - r_{max})$ divided by $(K - 0)$. Or, more simply, the slope is $-r_{max}/K$. Now replace m in our equation with $-r_{max}/K$, which gives

$$r_t = r_{max} + \left(\frac{-r_{max}}{K} \right) N_t. \tag{3.1}$$

Using algebra that you might have learned in high school, collect the two appearances of r_{max} and the result is

$$r_t = r_{max} (1 - N_t/K). \tag{3.2}$$

We did it! We have an equation that shows how per capita growth rate depends on population density in a linear way, like that exhibited for the black-throated blue warblers of Figure 3.3. Equations 3.1 and 3.2 are perfectly equivalent, but Equation 3.2 corresponds to an interesting ecological interpretation. Specifically, r_t is equal to r_{max} multiplied by a correction factor (i.e., $1 - N_t/K$). Think of N_t/K as an index of competition. When $N = 0$, there is no intraspecific competition, the index of competition (N_t/K) is equal to 0, and r_t is equal to r_{max}.

Conversely, when $N = K$, competition among individuals is greater, death rates are higher, and reproductive rates are lower. The competition index (N_t/K) is equal to 1, which drives r_t to 0.

When N exceeds K, then r_t will be less than zero and abundance will decline.

Let's go further and develop an equation that predicts next year's abundance (N_{t+1}) from this year's abundance (N_t) if the population's per capita growth rate is as described by Equation 3.2. To do so, we need to recall Equation 2.2 from the previous chapter, which I've reprinted here:

$$r_t = (N_{t+1} - N_t)/N_t.$$

Notice that the left side of this equation is identical to the left side of Equation 3.2. Also note that Equation 2.2 is always applicable to any population. It rises from the universal idea that populations are always well described by proportional growth.

Equation 3.2, however, is not always applicable; it applies only when the population dynamics are (linearly) density-dependent as are those exhibited by the population of black-throated blue warblers. Nevertheless, there is an equivalency between Equations 2.2 and 3.2, and that equivalency allows us to write a new equation that predicts abundance for a population with dynamics that are density-dependent. We'll set the left side of this new equation equal to the right side of Equation 2.2, and we'll set the right side of this new equation equal to the right side of Equation 3.2. The result is

$$(N_{t+1} - N_t)/N_t = r_{max} (1 - N_t/K).$$

To get an equation that predicts next year's abundance from this year's abundance, simply solve this new equation for N_{t+1}, which gives

$$N_{t+1} = N_t + N_t (r_{max} (1 - N_t/K)). \tag{3.3}$$

Compare Equation 3.3 and Equation 2.1 (from the previous chapter) for a subtle, but important similarity. To help with the comparison, here is Equation 2.1:

$$N_{t+1} = N_t + N_t (r_t).$$

Both equations are structured similarly, except that the per capita growth rate is represented by a single symbol (r_t) in Equation 2.1 and by a more complex expression ($r_{max} (1 - N_t/K)$) in Equation 3.3.

Think of it this way: Equation 2.1 says populations grow as a function of per capita growth rate (r_t), but Equation 2.1 doesn't offer any clue about how the value of r_t is expected to change over time. Equations 3.2 and 3.3 are more specific, describing cases where r_t tends to decline linearly with increases in N_t. Equations 3.2 and 3.3 are also sometimes referred to as *logistic population growth*.

With northern gannets we saw that population biology is connected to the ecological lives of individual organisms. The same is true with these warblers. When population density is greater, territory size declines and parents spend less time foraging and fledge fewer chicks. While it was not reported for this particular population of warblers from that forest in New Hampshire, other research has shown there is also a tendency for male warblers to spend more time singing to defend their territories as population density increases.

HOW STRONG IS DENSITY DEPENDENCE?

When the density of black-throated blue warblers is less than K, r_t tends to be positive, and density tends to increase. Conversely, when warbler density is greater than K, r_t tends to be negative and density tends to decrease. If the population were subject to no influence aside from the density-dependent processes represented by Equation 3.2, then population abundance would change until it reached K and then remain at K (Fig. 3.4).

This circumstance raises a question: For any particular population, what other influences might it be subject to, and how strong is density dependence relative to these other influences?

One important way to characterize the strength of density dependence is by the slope of the relationship between N and r, which is $-r_{max}/K$. When comparing the slopes of two populations, the population with the steeper slope is said to have stronger density dependence. Populations

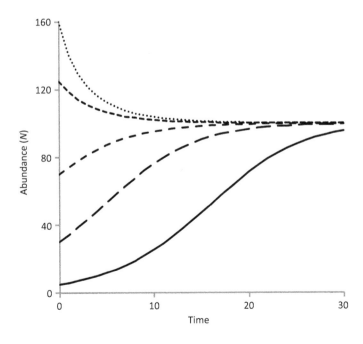

Figure 3.4 Negative density dependence is associated with dynamics referred to by mathematicians as *stable*. In this context, K is a *stable equilibrium*. No matter what the population's abundance is, abundance tends to go toward K. These lines were created in Excel using Equation 3.3 with $r_{max} = 0.2$ and $K = 100$. You should have a try at reproducing this graph. Time on the x axis may be measured in years or some other unit of time that corresponds to the rate at which a population's dynamics unfold.

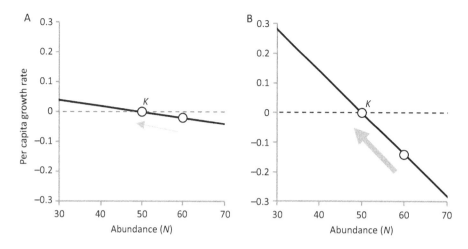

Figure 3.5 Weak (A) and strong (B) density dependence. The solid lines show the per capita growth rate (r) expected for given values of N for two populations. Suppose some favorable environmental condition in year t bumped each population from $N_t = 50$ to $N_{t+1} = 60$. At 60 individuals, the expected r for population A is only slightly negative (−0.02), but for population B, r is expected to be strongly negative r (−0.14). Both populations tend to return toward K, but the tendency is stronger for population B. Because the return tendency for A is weaker, some favorable environmental condition at time $t+1$ could overpower the influence of density dependence, causing the population to grow even farther from K.

with steeper slopes also tend to recover more quickly from perturbations, such as a year with extreme weather or excessive harvest (Fig. 3.5).

A second way to characterize the strength of density dependence can be appreciated by reconsidering panel B of Figure 3.3. The trend line captures the tendency for r to decline with increasing N, yet most of the field-based estimates of N and r do not fall right on the trend line. The statistical technique for calculating and assessing trend lines is called *regression*. That assessment includes the calculation of R^2, which may be used to judge the strength of density dependence. R^2 is a number between zero and one. If all the points from a data set fall exactly on the trend line, then $R^2 = 1$. If there is absolutely no trend in the data at all, then $R^2 = 0$. Using slightly more technical language, we can say that R^2 is the proportion of variance in the y variable (that's r_t for us) explained by the regression line. The remaining variance in the y variable is said to be unexplained (by the regression line).

For the warbler data, R^2 is 0.47. In other words, 47% of the variation in r_t can be predicted or explained by N_t. In more ecological terms, if Equation 3.2 is an appropriate description of the population, then 47% of the variation in r_t is attributable to the influence of various density-dependent processes. The remaining unexplained variation is attributable to other processes, such as errors in the estimation of N_t and r_t or fluctuations in environmental conditions (for instance, weather).

This example raises a broad question: Among studied populations, what portion of the variation in r tends to be explained by density dependence? Researchers from England and Finland offer a clue (Lindström et al. 1999). They analyzed 34 time series of abundances representing various species of birds, mammals, and insects. For each, they calculated the value of R^2 from the best-fitting trend line for each time series. From their analysis they found that R^2 ranged from 0.09 to 0.90. The median value of R^2 was 0.43, and the interquartile range was [0.26, 0.57]. In other words, one can readily find animal populations in which density explains virtually none (0.09) or virtually all (0.90) the fluctuations in population abundance. But for most populations, density explains between about one-quarter (0.26) and almost two-thirds (0.57) of the fluctuations in population abundance. Here, I'm using the words "fluctuations in population abundance" to stand in for r. While the Lindström et al. (1999) study is informative, a great deal more sophisticated research has been conducted on the strength of density dependence among populations. For a toehold onto that research, see the online supplement.

Now would be a good time to begin working with the practice problems in the online supplement.

NONLINEAR DENSITY DEPENDENCE

Bank voles are cute little fuzz balls that rest in burrows and feed from a pretty diverse menu including buds, seeds, fruits, and leaves. They don't live long—typically just 12 to 24 months—but they sure can reproduce. A female is quite capable of raising six pups in each of four litters during the summer. They experience intraspecific competition just like any species, but their short lives and fast reproduction rates mean that competition looks a little different from what we've seen with the warblers, who live a bit longer and reproduce a bit more slowly.

For bank voles, low abundance (N) means rapid reproduction. Furthermore, *when N is small*, a small increase in abundance results in females producing far fewer surviving pups, causing a dramatic decline in per capita population growth rate. By contrast, *when N is large*, a small increase in abundance leads to a reduction in r, but the reduction is modest because reproductive rates had already become low (as N increased from lower values of N) and survival rates (for a short-lived creature) are always low, regardless

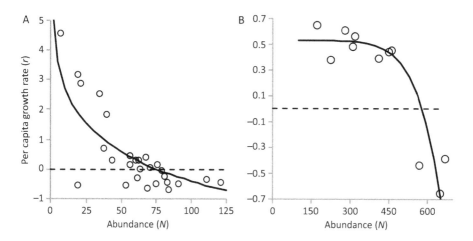

Figure 3.6 Nonlinear density dependence. Bank voles (A) and snail kites (B) illustrate a common circumstance by which density dependence is nonlinear. The nonlinearity is associated with the species' life-history strategies. Bank voles live shorter lives with higher rates of reproduction, while snail kites live longer with slower rates of reproduction. The trend lines are representations of Equation 3.4, which features the parameter θ. Shapes like A occur when θ < 1, and shapes like B occur when θ > 1. Panel A is based on data taken from the Global Population Dynamics Database (population ID 47). Panel B is based on Beissinger (1995).

of density. The result is a nonlinear relationship between N and r, where the slope is steep at low N and shallow at higher N (Fig. 3.6, panel A).

> Make sure you can identify where K is located in each panel of Figure 3.6. If not, you'll find a clue in Figure 3.5.

Now compare bank voles to snail kites, which are birds of prey that live in the Neotropics and feed almost entirely on, well, snails. Compared with bank voles, snail kites live longer—up to nine years of age—and reproduce more slowly—successfully fledging only two or three chicks each year.

For snail kites, *when N is small* and resources are abundant on a per capita basis, adults will survive to see another year and reproduce at their usual slow rate. Under those conditions (small N), a small increase in abundance causes rates of reproduction and recruitment to decline; but those rates were already low, so they cannot get too much lower. So, the overall effect is a modest decline in r. However, *when N is large*, small increases in N result in a big decline in adult survival, and because adult survival had been high, it can drop greatly.[1] When it does, the impact on r is a big decline. The result is a nonlinear relationship between N and r, where the slope is shallow at low N and steep at higher N (Fig. 3.6, panel B).

Your inner population biologist is whispering, "We need an equation that can describe these patterns that we're seeing." You are in luck, because others before us have found that a minor tweak to Equation 3.3 will do the trick:

$$r_t = r_{max} (1 - (N_t/K)^\theta). \tag{3.4}$$

Here the ratio N_t/K is raised to some power, represented by the symbol θ. Values of $\theta < 1$ give shapes like that shown in panel A of Figure 3.6, and values of $\theta > 1$ give shapes like that shown in panel B. When $\theta = 1$, Equation 3.4 becomes identical to Equation 3.3.

The values of θ, r_{max}, and K that best fit any particular data can be ascertained with more specialized regression techniques. We won't get into those more specialized techniques, but if we did, we'd learn that good estimates for the bank vole are $\theta = 0.091$, $r_{max} = 20.5$, $K = 66.5$, and for the snail kite they are $\theta = 3.3$, $r_{max} = 0.62$, $K = 543$. With those estimates and Equation 3.4 you can calculate r_t for any value of N_t.

(In Excel, you can perform the calculation indicated by Equation 3.4 for the snail kite in this way: "= 0.62*(1 − (A1/543) ^ 3.3)" if cell A1 contains the population's abundance.)

Population dynamics may depend importantly on whether and how density dependence is nonlinear. For example, populations with $\theta > 1$ (as in Fig. 3.6, panel B) have a stronger tendency to fluctuate close to K, because density dependence is strong near K. To better see this, compare panel B of Figure 3.6 (at the part of that graph where N is close to K) with Figure 3.5.

By contrast, populations with $\theta < 1$ (as in Fig. 3.6, panel A) often wander far from K. This is because the slope of the trend line near K is shallow, which means density dependence is not a strong force pressing a population back toward K. These populations will sometimes be far from K (above or below).

Finally, while a population may be nonlinearly density-dependent, that nonlinearity may not be apparent unless you have observed the population when N is close to zero. For example, suppose that you never observed the data points for $N < 30$ in panel A of Figure 3.6. In that case, the relationship would look linear, even though it is actually nonlinear over the full range of N.

ENVIRONMENTAL STOCHASTICITY

Alpine ibex were introduced to Switzerland's national park in 1920. Their population dynamics exhibit a shape that is classically associated with density-dependent growth (compare Fig. 3.7, panel A with the solid line

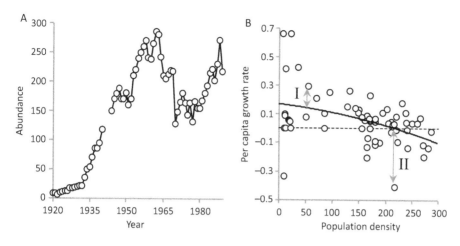

Figure 3.7 Temporal fluctuations in abundance for a population of ibex reintroduced to Switzerland's National Park in 1920 (panel A). Panel B shows that density dependence is slightly nonlinear, as indicated by the solid trend line. The trend line has an R^2 value of 0.16, which means it explains 16% of the variance in r. Panel B also highlights two residuals, labeled I and II (see text for details). Statistical analysis of the residuals indicates that about 25% of the variance in r is attributable to year-to-year fluctuations in winter precipitation. Adapted from Sæther et al. (2002).

in Fig. 3.4), where growth is comprised of what might be called a growth phase while abundance is low, followed by a phase where abundance fluctuates around what seems to be constant level.

The plot of N versus r exhibits a relatively weak nonlinear trend obscured by much fluctuation (Fig. 3.7, panel B). Given the equations that we've discussed thus far, the best-fitting model for this population's dynamics is $r_t = 0.14 \times (1 - (N_t/212)^{1.54})$. That model explains about 16% of the variance in r_t (i.e., $R^2 = 0.16$).

The unexplained variance is represented by what statisticians refer to as *residuals*, which are a series of numbers representing differences between the field-based estimates of r_t (observed data points) and the predicted values of r_t (which lie on the trend line). Each observed data point has its own residual (Fig. 3.7, panel B). Positive residuals are above the line and negative residuals are below the line.

Residuals allow one to pursue the questions of why some observations are greater than expected (positive residuals) and why other observations are lower than expected. Or, what else aside from density dependence is occurring that can account for why some residuals are greater during some years and lesser during other years.

For this ibex population, researchers showed that negative residuals tended to occur during years with more winter precipitation (Sæther et al. 2002). In other words, more winter precipitation tends to reduce population

Figure 3.8 An ibex in their foraging environment. Source: https://upload.wikimedia
.org/wikipedia/commons/b/bd/Alpine_ibex_Cima_di_Terrarossa_3.jpg.

growth. More precisely, winter precipitation accounts for about 25% of the variance in r_t. So, 16% of the variance is due to the influence of density-dependent processes (such as intraspecific competition), and 25% of the variance in r_t is due to interannual fluctuations in winter precipitation.

The influence of winter precipitation on ibex population dynamics also corresponds to the ecology of ibex. They feed on grasses at higher elevations above the tree line (Fig. 3.8). During years with more snow, there are fewer snow-free patches in late winter and early spring, which means that there is a decline in the availability of food at that time of year, and consequently, a decline in population growth.

In this example, winter precipitation may be considered a kind of *environmental stochasticity*, where an approximate synonym for "stochasticity" is "randomness" or "chance." Think of environmental stochasticity as aspects of the environment that influence *r*, for which the direction and magnitude of that influence varies over time in a more-or-less unpredictable manner. Year-to-year fluctuations in various aspects of weather often count as environmental stochasticity.

POSITIVE DENSITY DEPENDENCE

While negative density dependence is a common circumstance, there are important instances in which *r* tends to increase as *N* increases. This is called *positive density dependence*.

This dynamic is illustrated by Vancouver Island marmots, which are highly endangered and endemic to the island of Vancouver on Canada's Pacific coast. Marmots are large rodents that live in high alpine meadows where they feed on grasses and phlox in the spring, move onto lupines and other forbs as the summer progresses, and then hibernate when the sun is finished providing the year's supply of greens.

They burrow below ground to rest and for protection from predators, which include golden eagles, wolves, and mountain lions. Marmots live in colonies and are intensely social, typically spending 90% of their time within 20 meters of other colony members.

Vancouver Island marmots have been threatened by loss of habitat, mostly in the form of clear-cutting the lower-elevation forests that separate alpine meadows. Young adult marmots disperse from their colonies in search of mates and need to cross those forests to do so. When the forests are clear-cut, they are vulnerable to predation.

By the year 2003, fewer than 30 of these marmots were living in the wild. As the population declined, its capacity to recover was further diminished by what might be called a breakdown in the benefits of sociality.[2] While foraging in their alpine meadows, marmots spend some time being on the lookout for predators. When a marmot spots a predator, they let loose a warning cry, and all the others run for their burrows. When the colony is large, each individual devotes a little time and effort to being vigilant. As colony size decreases, each individual's effort to be vigilant increases, and that, in turn, reduces their rate of foraging.

During the 2000s, when the population was especially low, marmots spent 10 times more time being vigilant and experienced an 86% decline in feeding rate, as compared with the 1970s when there had been more marmots. Whereas marmots are normally most active in the morning and early evening, during the 2000s, this pattern of activity was largely lost.

Additionally, during the 2000s male marmots and lone marmots tended to have home ranges that were 10 to 60 times larger than normal—presumably in response to the difficulty of finding colony mates or mates with whom to reproduce.

As the marmot population fell to low levels, individual marmots were living essentially solitary lives, interacting with other marmots only 10% of the time that might be considered normal.

These behavioral changes for individual marmots had big implications at the population level. As population size (and colony size) got smaller, there was a tendency for the subsequent per capita growth rate to also decline. In other words, the population exhibited positive density dependence, as illustrated in Figure 3.9.

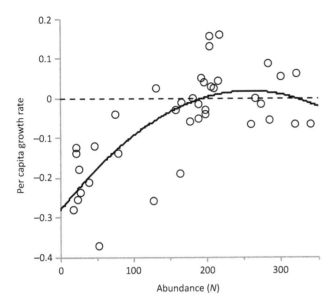

Figure 3.9 Per capita growth rate in relationship to abundance of Vancouver Island marmots. This is an example of positive density dependence. Reductions in abundance tend to be followed by further reductions in abundance whenever N is less than about 250. That is, the trendline is positive for N < 250. Adapted from Brashares, Werner, and Sinclair (2010).

In recent years, thanks to herculean efforts by conservationists, the marmots of Vancouver Island are doing better. By 2014, more than 200 marmots were living in the wild.

A second example of positive density dependence involves a population of bighorn sheep living in the foothills of the Canadian Rockies just west of Calgary.[3] In a typical year no sheep were killed by cougars, who normally fed on elk. But in some years, an individual cougar or two would make a habit of killing sheep in this population.

In Chapter 9, we'll learn how the impact of predation tends to increase as the abundance of prey declines. For now, you'll get the idea from this hypothetical situation: if a cougar kills 1 sheep in a population of 100, then the mortality rate due to predation is 1%. If that same cougar kills 1 sheep in a population of 20, then the rate of predation mortality is 5 times greater, at 5%. The point is that the impact of predation depends not so much on the *number* as the *proportion* of the population that dies from predation.

Something like that hypothetical situation occurred between these cougars and sheep. Every year some more-or-less constant number of sheep were preyed on, and that number corresponded to an increasing proportion of sheep killed as sheep abundance declined. That dynamic was most apparent in the rate at which yearlings were recruited into the population such that higher rates of predation corresponded to lower rates of recruitment (due to higher predation mortality among young sheep). As a result, recruitment rate was positively density-dependent (Fig. 3.10, panel A).

We started this chapter supposing that negative density dependence is due entirely to intraspecific competition, but a better way to think of it is that

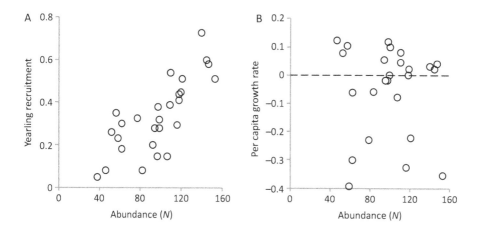

Figure 3.10 Panel A shows that recruitment of yearling sheep tended to be lower when the abundance was lower. If recruitment was primarily influenced by intraspecific competition, then one would expect a negative trend. This positive trend is a result of cougar predation (see text for details). Yearling recruitment is one of several vital rates that contribute to the overall per capita growth rate. Some vital rates may be negatively density dependent and others positively (as in panel A). Panel B shows how for this population the combined influence of all vital rates on r results in no real trend between N and r. The dynamics are essentially density-independent. Adapted from Bourbeau-Lemieux et al. (2011).

populations are simultaneously influenced by many processes that involve density, such as competition, predation, disease, and sociality. A graph showing the relationship between N and r is the combined result of all those processes. For example, intraspecific competition in a sheep population might result in negative density dependence, so long as the sheep are not exposed to predation. And predation might have an influence that is—by itself—positively density dependent. When predation's "positive" influence is combined with competition's "negative" influence, the result can be weak density dependence, where the relationship between N and r is not readily detectable, as in panel B of Figure 3.10.

A population can have both positive density dependence (when N is smaller) and negative density dependence (when N is larger). For example, the marmot population has positive density dependence for $N < 250$ and looks to have negative density dependence for $N > 250$ (Fig. 3.9). The resulting dynamics are interesting and complicated and the subject of an exercise in the online supplement.

DELAYED DENSITY DEPENDENCE

Earlier in this chapter, we heard about bank voles to understand nonlinear density dependence. They are one of about 150 species of vole. Now let's

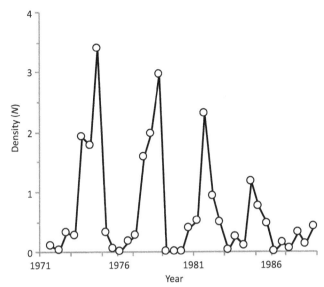

Figure 3.11 Cyclic fluctuations in a population of gray-sided voles. The y axis is an index of density, specifically the number of voles captured for every 100 trap-nights of effort. Vole density was estimated each summer and autumn. Adapted from Hörnfeldt (1994).

look at some gray-sided voles. Both species live across Europe and northern Asia. Bank voles are smaller and quicker than gray-sided voles. Bank voles also have a more diverse diet compared with gray-sided voles, who are especially partial to feeding on the leaves of dwarf *Vaccinium* shrubs (same genus as blueberries).[4]

Figure 3.11 features the dynamics of a population of gray-sided voles living in northern Sweden. The population exhibits short bursts of impressive growth, followed by equally impressive crashes. The population peaks are more than 10 times greater than the population lows. More impressively, that pattern repeated itself like clockwork every three or four years—at least until the late 1980s when the population cycles fizzled.

Before we discover the cause of these population dynamics, let me give you a behind-the-scenes look at how the data in Figure 3.11 were found. Birger Hörnfeldt (1994) and his research team identified a study area of mostly managed forest in northern Sweden. This study area was 60 miles by 60 miles. Within it, Hörnfeldt identified 58 regularly spaced plots. Each plot was 1 hectare. Fifty traps were located at regular spacing within each 1-hectare area. This is a total of 2,900 traps. Each trap was tended for three nights each spring (May) and each fall (September). The researchers did this for 17 years (1971–1988)!

In total, Hörnfeldt's team trapped 7,891 bank voles, 2,010 gray-sided voles, and 1,455 field voles. These represent more than 90% of all the small mammals that were trapped. The other individuals were various species of mouse or shrew.

(Hörnfeldt did not report whether the animals were live-trapped, marked, and released, or snap-trapped. Today, it would be common to live-trap.)

Hörnfeldt tallied the total number of each vole species trapped each season across the study area. This number of trapped individuals was divided by the number of trap-nights of effort each season to result in an index of abundance. The result for gray-sided voles is shown in Figure 3.11.

Before I tell you how Hörnfeldt analyzed the data, we need to revisit and ever-so-slightly modify the equation that we've been using to represent density dependence. Draw your attention back to Equation 3.3, specifically to that index of competition, N_t/K. Recall that decreasing levels of competition are represented by smaller values.

For Equation 3.3 to provide a good representation of a real population, certain processes have to play out very quickly. There is an increase in N, then an increase in competition (think, N/K), then reproductive and survival rates decline—all three process have to occur more-or-less immediately if Equation 3.3 is to be a good representation of the population. Often enough, Equation 3.3 (or Equation 3.4) is a good representation (e.g., Fig. 3.2, panel B, Fig. 3.3, and Fig. 3.7).

But sometimes there is a delay in that chain of events. N increases, and competition intensifies, but animals don't starve to death right away. There might be reductions in reproduction that are not manifest until after enough time passes for the reduction in food to take its toll. It might take a year or so. If so, then this year's growth rate (r_t) might be better indicated by N during a prior year, so Equation 3.3 can be rewritten with a modified index of competition:

$$r_t = r_{max} \left(1 - N_{t-T}/K\right), \qquad (3.5)$$

where T can stand for some relatively small positive integer, say 1 or 2 years. In other words, this year's growth rate is influenced by population density from sometime in the past.

Whereas Equation 3.2 tends to result in relatively simple dynamics (Fig. 3.4), the dynamics of Equation 3.5 can be more complicated. In particular, if r_{max} is large enough, then Equation 3.5 predicts a tendency for the population to overshoot K before declining (Fig. 3.12). As the population declines, it undershoots K, and then increases again. Each oscillation is smaller than the previous, and eventually N settles at K. In real populations, environmental stochasticity would be present and prevent the population from ever settling at K. Such populations might cycle in perpetuity or at least exhibit persistent fluctuations reminiscent of cycling. How remarkable

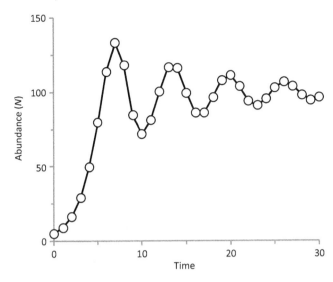

Figure 3.12 A population with delayed density dependence, which leads to dampened oscillations around K. This time series was built using Equation 3.5, setting $r_{max} = 0.85$, $K = 100$, $T = 1$, and $N_0 = 5$.

that such a small adjustment to the equation for density dependence can lead to such complicated dynamics.

Now, back to the gray-sided voles. With the data from Figure 3.11, one can make a graph to see how growth rate during the summer (r_t) is related to population density at the beginning of the summer (N_t). The result is panel A of Figure 3.13, which gives the impression that density explains little variance in r. The population appears nearly density-independent. But when we compare growth rate during the summer to abundance at an earlier time (the previous fall), we find a stronger relationship. This is the influence of *delayed density dependence* (Fig. 3.13, panel B).

These gray-sided voles from Sweden represent an idea that extends far beyond their little furry selves. About 30% of all studied populations seem to cycle (Kendall, Prendergast, and Bjørnstad 1998), and population biologists have long wondered why. Delayed density dependence is now understood to be an important part of the answer.

The next curious observation is that populations of many small mammals living at high northern latitudes cycle, but populations of the same species living farther to the south tend not to cycle. What's up with that?

The answer seems to lie with predation. The northern populations tend to be preyed on by specialist predators (especially weasels) for whom the small mammals are the predators' only source of food. But the southern populations are preyed on by generalist predators (such as foxes), who are nourished by a diversity of prey (including, for example, small birds).

The population dynamics of specialist predators are tightly linked to the dynamics of their prey. When their prey declines in abundance, soon too

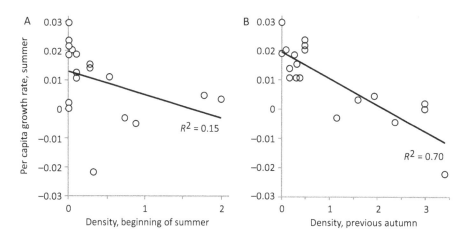

Figure 3.13 Per capita population growth rate during summer (*r*) for the vole population depicted in Figure 3.11. Growth rate is shown in relationship to density at two different times—the beginning of summer (A) and the prior autumn (B). Delayed density dependence (B) is a better predictor than the relationship with no delay (A). Such delays are an important explanation for population cycling. Adapted from Hörnfeldt (1994).

will the predators. But the linkage for generalist predators is not so tight. When the abundance of small mammals declines at a southern locale, the foxes can switch to different kinds of prey. So a decline in voles does not have to be linked to a decline in foxes.

It turns out that specialized predators are the source of delayed density dependence in cycling populations of small mammals. What happens is this: vole abundance increases (because predator abundance is low), and this leads to an increase in predators, which will slow population growth of the voles. But it takes a little time (six months to a year) for the predator population to increase in response to increased voles. That's where the delay comes from.

Finally, look at the right side of Figure 3.11, where cycling in the vole population appears to have fizzled during the mid-1980s. Similar fizzling of population cycles was observed throughout Europe and Fennoscandia during the 1980s and 1990s in several species of vole, black grouse, and most impressively in larch budmoth in the European Alps. The loss of cycles with budmoth is impressive because dendrochronology indicates it to be the first time in 1,200 years that budmoth cycles had ever been lost. At least some of these populations may have since regained their cycling, but the cause and full extent of the loss (and recovery) remain a bit of a mystery. Climate change is suspected.

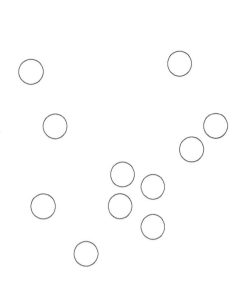

CHAPTER 4

Ethical Dimensions

Whooping cranes have been steadily increasing for decades. Perhaps some of the special protections afforded to whooping cranes *should* be rescinded—even though it may be a century before they are restored to their former abundance? *Should* the catch of deep-sea fish be limited, even if doing so diminishes the well-being of some commercial fishers? *Should* the elephants of Kruger National Park be culled when they become overabundant?

Every example from Chapter 2 and many ideas from Chapter 3 involve some question about what should be done. This is not surprising because the conservation of animal populations is fundamentally about how we should treat them. Furthermore, anytime we contemplate what we should do, we are contemplating ethics. For those simple reasons, conservation is ethics in action. The connection between conservation and ethics is also inescapable. One may neglect the ethical aspects of conservation, but doing so does not make it go away. Neglect only increases the chances of mishandling those ethical aspects.

Because we use the word "ethics" in our everyday language to refer to various ideas, it's important to clarify its meaning here. "Ethics"—as the term is used in this book—is *not* telling others what to do. It can be fine to tell others what to do, but that's not the central concern of ethics. Rather, ethics is an academic discipline that provides formal assessment of claims about what we should and should not do. The purpose of ethics is to better understand and better explain how we should act.

Ethical inquiry is a formal method for evaluating claims such as "We should continue protecting whooping cranes." Evaluating the appropriateness of any ethical claim (including claims about conservation) requires

- identifying which facts and values are relevant to the claim
- synthesizing those facts and values into insight about the appropriateness of the claim.

Some values are simple and seemingly easy to evaluate, such as "Preventing extinction is good." Other values are complicated, such as whooping crane protections are important enough to merit placing some restrictions on how some landowners use their land. The relevant facts will typically include scientific knowledge about animal populations, such as whooping crane numbers have been growing, on average, by ~4.4% per year from the 1950s to at least the 2010s.

> Ignoring the ethical dimension of conservation risks mishandling it. To properly handle the ethical dimension of conservation requires a method of thinking that synthesizes facts and values.

You'll soon see that scientific facts play a complicated role in ethical inquiries about conservation. Understanding that complexity is important for better understanding difficult decisions about the conservation of animal populations. So, the subject of this chapter is important because gaining scientific knowledge about animal populations without learning methods for evaluating its application would be like building a ship without including a rudder or compass.

ARGUMENT ANALYSIS

The essential tool for synthesizing facts and values is argument analysis. Here, an argument is not a fight or disagreement. Rather, it is a formal method for thinking through an idea. Here's how it works. An argument is a conclusion, preceded by a set of premises that represent evidence necessary to support the conclusion. A simple, illustrative example is shown in Figure 4.1.

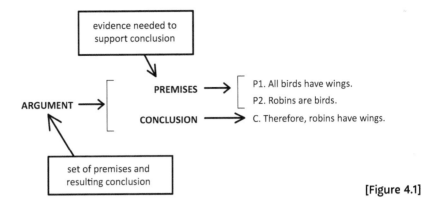

[Figure 4.1]

The argument is on the right, and the parts of the argument are labeled and described on the left. P1 and P2 are abbreviations for "premise 1" and "premise 2." C stands for "conclusion."

While Figure 4.1 is a good example of an argument, it is not an example of a special class of arguments known as *ethical arguments*. Ethical arguments are characterized by being able to express their conclusions using the word "should." An illustrative example is shown in Figure 4.2.

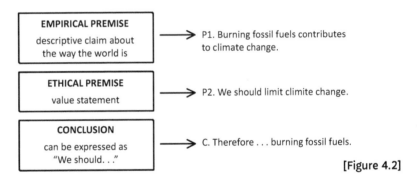

[Figure 4.2]

An ethical argument is always comprised of two kinds of premises, empirical premises that describe the world and ethical premises that represent some kind of value. This distinctive composition is how scientific facts and values are synthesized, resulting in insight about how we should act.

Arguments are useful because of the method by which they are analyzed. Specifically, an argument's conclusion is reliable if two conditions hold: (1) all of the premises are true, and (2) the argument's logical structure is valid. Some examples will help us understand. An argument to illustrate condition (1):

P1. All mammals give birth to live young.
P2. The duck-billed platypus lays eggs.
C. Therefore, the duck-billed platypus is not a mammal.

Because premise 1 is false, the conclusion is not supported by the argument. Strictly speaking, a reliable conclusion requires that all of the premises to be true. If even one premise in the argument is wrong, then the entire argument fails, and the conclusion is not supported. While the principle is simple, we will see that evaluating premises can be tough.

Here is an argument to illustrate condition (2):

P1. All lions are animals.
P2. All felids are animals.
C. Therefore, all lions are felids.

You recognize that the conclusion is true. Less obviously, the logical structure of the argument is invalid. The invalid logic is easier to recognize if we change the subjects and predicates of the premises, but retain precisely the same logical structure:

P1. Tadpoles are cold-blooded.
P2. Fish are cold-blooded.
C. Therefore, tadpoles are fish.

The falsity of the conclusion makes it easier to see that the logic is poor. Keep in mind that the logic is poor in exactly the same way in the felid argument.

The fish and felid arguments illustrate several important points about argument analysis. First, invalid logic is sometimes difficult to recognize, especially if you believe the conclusion is true (as in the felid argument). Second, an invalid argument does not guarantee that the conclusion is false; it only means that the conclusion is not supported by that argument. There may be some other argument that is sound and valid and that leads to the same conclusion.

Third, some invalid logical forms show up in real-world thinking so frequently that logicians have built a catalogue of logical fallacies to help non-logicians like us spot instances of invalid logic. For example, the felid argument makes use of a fallacious logical structure known as *fallacy of the undistributed middle*. Its formal structure is:

P1. X is A.
P2. Y is A.
C. Therefore X is Y.

Any argument with that logical structure is invalid, and its conclusion is unreliable.[1] If you type "logical fallacies" into an internet search engine, you'll find lots of good information about them.

In the meantime, here is a general strategy that can often (not always) reveal logical shortcomings of an argument. When studying an argument, ask this pair of questions: Is the argument missing a relevant premise, and is that missing premise false or inappropriate? Again, an example:

P1. We should promote the economy.
P2. Abusive child labor would promote the economy.
C. Therefore, we should promote abusive child labor.

An advocate for the argument might think it is perfectly fine. But a critic familiar with argument analysis would say the argument is missing a premise:

P3. Any method for promoting the economy is acceptable.

Premise 3 is necessary for the logic to be valid, but P3 happens to be an inappropriate premise. So, the argument fails, and the conclusion is not supported. The original argument (P1, P2, C) was invalid. Adding P3 helps us see it as such.

Before closing this section, I have two notes about the vocabulary we're starting to use. First, if the logic of an argument is poor, then the logician would say the argument is *invalid*. Second, we tend to classify empirical premises as true or false, and ethical premises as appropriate or inappropriate.

Some Perspective

The preceding examples give an idea of how arguments work, and we'll apply those ideas to the conservation of populations in a moment. But first, a couple of broad perspectives.

First, argument analysis is not only important for ethical inquiry, but also as an essential aspect of *critical thinking*. This connection merits attention because critical thinking is sometimes confused with thinking that levies criticism or thinking really hard. Yes, critical thinking often involves criticism and hard thinking, but that is not what critical thinking is. Rather, critical thinking is the robust analysis.

The second perspective involves a common misperception about ethical inquiry, which is that ethics and values are like chocolate cake in the sense that you might like chocolate cake, but I do not, and that's all there is. Nothing further to say, and no basis whatsoever for assessing whether one view is more appropriate than another. That view of ethics and values is far from the truth. Ethical claims require reasons, and argument analysis is key to evaluating the quality of those reasons. This does not preclude legitimate disagreement between people about how we should behave. But it does mean that ethics is not simply an announced preference that requires no further reason or explanation, such as "I like chocolate cake."

Culling Elephants and Ecosystem Health

We have seen in Chapter 2 that elephant abundance in Kruger National Park grew exponentially throughout much of the twentieth century (Fig. 2.2). In 1967 the park was inhabited by more than 6,500 elephants. Wildlife managers observed the impact of elephants on Kruger's plant community, judged the park's carrying capacity of elephants to be 7,000 (~1 per square mile), and began culling elephants. Culling was conducted annually for nearly three decades, until 1995. By that time, public pressure from across the world led to the end of elephant culling in Kruger. The government of South Africa issued a new management plan in 1999 whose position was summarized by Dickson and Adams (2009, p. 113) this way: "Fluctuations in elephant populations were broadly beneficial for biodiversity, but an unmanaged population would eventually trigger negative environmental impacts (Whyte et al. 1999)." Consequently, the new plan advocated culling elephants if nonlethal methods of managing elephants were deemed, according to Dickson and Adams (2009, p. 113), to be "'inadequate, unfeasible or inappropriate' (Whyte 2004, p. 102)." The plan triggered years of intense debate, but culling never resumed. By 2019, elephant abundance had grown to nearly 20,000 (Fig. 4.3).

Outside Kruger National Park, elephants are killed for various reasons—legally and illegally. Aside from poaching elephants for their ivory and providing wealthy trophy hunters with opportunities to kill exotic creatures, the primary reasons that humans kill elephants are in response to crop damage, threats to human safety, and threats to the ecosystem health of protected areas. A concern in protected areas is that high elephant density tends to reduce tree density, which in turn leads to more open, savanna-like environments. Those changes lead to other changes in the plant community, including the reduction of plant biodiversity. This had been the concern in Kruger.

Decisions about killing elephants to protect crops and ecosystem health tend to be complicated and controversial with influential advocates taking various positions. What I'll do next is use elephant culling for the pur-

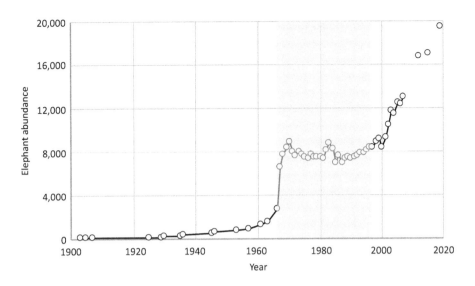

Figure 4.3 Abundance of elephants in Kruger National Park, 1903–2019. The gray-shaded area indicates the period of culling, 1967–1995. During that time over 14,000 elephants were shot, and almost 1,500 elephants were forcibly relocated—mostly orphaned juveniles. The culling and relocation represented the removal of approximately 7% of the population each year. Data sources: Whyte, van Aarde, and Pimm (2003), Whyte (2007), Ferreira, Greaver, and Simms (2017), and Hendry (2021).

pose of protecting ecosystem health as a case to show how ethical arguments may be used to synthesize scientific facts and values and to assess complicated issues.

Building an Argument

When it comes to killing elephants or any other ethical concern, your intuition may powerfully lead you to favor one view or another. That is normal and often a good thing. Nevertheless, an important stance for evaluating an ethical claim is to temporarily suspend your belief (as best you can) and analyze the ethical argument as if nothing were at stake, as if you were solving a crossword or putting a puzzle together. This disposition is valuable for discovering insight that you may not already have.

Think of it this way: the point of building and analyzing an ethical argument is to develop a deeper understanding of what's right or wrong. If your mind is already firmly set, if you are already completely convinced that something is wrong or right, then that mindset can be a formidable obstacle to developing new insight. It's good to hold firm ethical beliefs. I'm just saying that gaining certain insight—such as why other people believe differently than you—is less likely while being held by firm conviction. Let's give it a try.

In a state of open-mindedness, let's build an argument about elephant culling. The first step is to state the conclusion:

C.   Elephants should be culled in protected areas when they are overabundant.

We start with the conclusion not because we do or do not believe it. We start here because it's a claim we want to evaluate.[2]

Now we think of premises that would have to be true to support that conclusion. We can channel the mind of an imaginary advocate for culling, who'd likely suggest:

P1.   Overabundant elephants harm ecosystem health.
C.   Therefore, elephants should be culled in protected areas when they are overabundant.

Yes, it sure seems we'd need P1 in an argument that supports this conclusion. But that's not enough. This is an ethical argument, and ethical arguments need at least one ethical premise. We can help our imaginary advocate express their idea more completely by suggesting an additional premise:

P1.   Overabundant elephants harm ecosystem health.
P2.   We should take actions that protect ecosystem health.
C.   Elephants should be culled in protected areas when they are overabundant.

While P2 may seem modest and unimaginative, it is typically better to be explicit about the ethical motivations. Soon, we'll see that P2 reveals worthwhile insight.

Imagine a second advocate who is opposed to culling elephants. Suppose they concede the relevance of P1 and P2, but also express concern that the argument is missing something. Specifically, a claim about right ways to treat individual elephants. To account for that concern, we can add a premise:

P1.   Overabundant elephants harm ecosystem health.
P2.   We should take actions that protect ecosystem health.
P3.   Culling is morally acceptable treatment of elephants.
C.   Elephants should be culled in protected areas when they are overabundant.

Our imaginary advocate (the one who objects to the conclusion) might exclaim that P3 is wrong. As argument analysts, our response should be: at this moment we're building an argument that contains all the premises that *would have to be true (or appropriate) if* the conclusion is to be supported. Evaluating which premises are true (or appropriate)—that is a separate step that comes later. For now, let's just acknowledge that *if* the argument supports this conclusion, then P3 would likely have to be included *and* judged appropriate. If not, then the argument would fail.

Suppose that both advocates are satisfied that this three-premise argument includes all the key premises. If so, then we'd seem ready for the next step, which is to evaluate the appropriateness of each premise. But we're not there yet. I see another missing premise that is likely to be important and thus further our thinking about the issue. Specifically, let's add a premise and place it between P2 and P3, as they are numbered above. Doing so will cause us to renumber the premises, yielding:

P1. We should protect ecosystem health.
P2. Overabundant elephants harm ecosystem health.
P3. The plan for killing elephants adequately limits harm to ecosystem health and does so with the least harm.
P4. Culling is morally acceptable treatment of elephants.
C. Elephants should be culled in protected areas when they are overabundant.

Strictly speaking the order of premises does not matter. Our newly labeled P3 could have been placed anywhere. Nevertheless, an argument is sometimes easier to follow if some attention is given to the ordering of the premises.

Suppose—just provisionally—that we have identified all the important premises. By doing so, we are supposing that the conclusion is connected to the premises by valid logic. If further analysis reveals the need to add another premise, we'll do so. Now we can move to the second step of argument analysis, which is to evaluate the truth or appropriateness of each premise. In doing so, we want to be genuinely critical—not dismissive, mean, or pedantic, but definitely critical. You don't want to be overly critical of premises that conflict with your intuition, or soft on premises that align with your intuition.

The risk of being soft on premises that agree with your intuition is likely greater than is obvious. A well-studied phenomenon in psychology is the human tendency to more readily see thoughts that support one's intuition and overlook or prematurely dismiss ideas that conflict with it. This powerful tendency is referred to as *confirmation bias*. If you are human, then you are vulnerable to this bias. Confirmation bias is also an obstacle to performing rigorous argument analysis. We've already been trying to limit confirmation bias by (1) making an active, conscious effort to temporarily suspend belief about an issue and (2) working with—or at least vividly imagine working with—those who'd have different views. In other words, discovering and evaluating all the relevant premises are greatly aided by contributions from people with different views.

Evaluating Premises on a First Pass

Observe that P2 and P3 seem to be scientific premises. As such, evaluating the truth of P2 and P3 would benefit from the knowledge of population biology. Furthermore, P1 and P4 seem to be ethical premises. So far, so good, because we know that ethical conclusions require facts and values—that is, empirical premises and ethical premises.

While each premise should be thoroughly evaluated, there is value in first performing a quicker, preliminary evaluation of the premises to get a feel for the breadth of considerations that are likely to arise. This initial pass often includes identifying key phrases that require clarification so that everyone reading the argument has the same understanding of its meaning.

For example, P1 and P2 both refer to "ecosystem health." To some, that phrase refers to the extent that an ecosystem provides some human interest. To others, that phrase refers to the extent that an ecosystem is not impacted by humans.[3] Evaluating the appropriateness of P1 and truth of P2 likely depends on clarifying the meaning of the phrase "ecosystem health." In some cases, such clarification can be achieved by providing a definition and then proceeding with the evaluation.

Premise 2 includes another idea that likely requires specification. That is, "overabundant." We'll need to answer the question, Overabundant in what sense? That is, abundant to the point of causing what problem? Without such specification, "overabundant" refers to misplaced concern for the number of elephants per se, rather than concern for some problem that elephants might be causing.

Clarifying "ecosystem health" may be had by revising P1 such that it reads:

P1′. We should ~~protect ecosystem health~~ maintain tree density and plant diversity in protected areas.

And to clarify "overabundant," we could replace P2 with

P2′. When elephant density exceeds x per square kilometer, then tree density and plant diversity tend to decline.

The letter x stands for some number that can be estimated with scientific research. Estimating that number may take time—even years of research. Until that research is conducted, scientists may not agree on a value of x, or they could largely agree to a wrong value. Moreover, x is not likely to be a fixed number. Rather, it's likely to vary with rainfall and perhaps the ecosystem's fire history. We can confront those challenges in time. For now, it serves to simply acknowledge that those concerns about x may arise.

Someone with more expertise about elephant ecology than we have might think P1′ and P2′ are inadequate representations of the concern. If so, the revisions we proposed will have been valuable for prompting further revision.

In any case, the kind of specificity that we're aiming for with P1′ and P2′ is liable to be important for evaluating the truth and appropriateness of those premises and consequently the appropriateness of elephant culling in protected areas.

While P1 and P2 (and their more specific expressions, P1′ and P2′) represent overarching goals or motivation for culling, P3 and P4 represent the methods by which the goal might be realized. And here it's important to note that ethics is as much about end goals as it is about means for realizing them. P3 and P4 raise a number of significant questions:

· How many elephants would be killed and over what time frame?
· If that number of elephants were killed, to what extent would the goal (as represented by P1′) be realized?
· To what extent can the same outcome be realized by some other method that involves less harm, insofar as killing elephants is a harm?

These considerations may give us occasion to revise P4—namely, that culling is morally acceptable treatment of elephants. After all, some might say that killing an elephant is always wrong—no matter what. If that is not the case, then the better expression of P4 is probably:

P4′. The harm caused by killing elephants is outweighed by the benefits of protecting ecosystem health, as described in premise P1′.[4]

The purpose of what we have done in this section is to identify the breadth of issues that are likely to arise in confronting the argument's conclusion.

It may seem daunting if there are numerous weighty issues, but that is the nature of complicated ethical judgments. In any case, once we have identified the breadth of issues, we can focus on each individual issue as closely as necessary. If new issues arise along the way, they can be represented by the addition of new premises. If all the issues can be resolved to the point of supporting each premise, then the argument will end up supporting the conclusion. If not, then it won't.

Finally, it is useful to think of ethical arguments as being a little like mathematical models in population biology. In both cases, their value lies not in being perfect, complete representations of the world. Population models are valuable when they contain just enough detail to provide genuine insight about how real populations work. If some ecological insight requires adding a bit of detail to the model, then it should be added, but not otherwise. Similarly, ethical arguments do not need to include every imaginably true premise. Rather, they need to include enough premises to reveal and develop insights that might otherwise be overlooked.

BROAD THEMES

Our goal in this chapter is not to resolve concerns about elephant culling, but to begin working with a method for addressing ethical inquiries. In the next several sections, we'll revisit some of the issues that arise in evaluating the premises of the elephant culling argument. This time, we'll probe each issue a little deeper and do so, in part, for the purpose of illustrating how these issues are not special to elephant culling in Kruger. Rather they represent broad themes in conservation ethics.

Overabundance

When elephant culling occurred in Kruger, it was justified by a pair of ideas similar to P1 and P2, except that the language included references to the carrying capacity (K) of elephants. More precisely, the factual claim had been that elephant abundance in Kruger NP had exceeded the carrying capacity of 7,000 elephants, and the value claim was that it is bad to exceed carrying capacity. Both ideas are almost certainly wrong.

If carrying capacity were 7,000 elephants, then abundance would have tended to decline when $N > 7,000$. This has not been the case, as evident from Figure 4.3. Furthermore, exceeding K is not reason to think elephants had become detrimentally overabundant (see Owen-Smith et al. 2006). It is normal and common for populations to spend considerable time above carrying capacity, often far above K (Fig. 4.4).

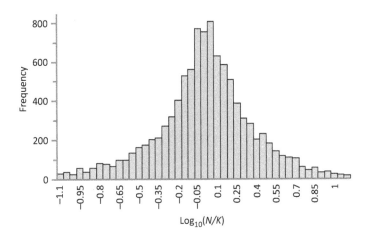

Figure 4.4 Abundance data compiled from more than 600 animal populations showing the frequency (*y* axis) that populations spend at different levels of abundance, in relationship to the carrying capacity (*K*). Populations spend most time near carrying capacity—that is, near zero on the *x* axis of this graph. Populations also spend slightly more than half of the time above *K*—that is, to the right of zero on the *x* axis. Adapted from Sibly et al. (2007).

Wildlife management that is too focused on controlling some prescribed *number* of animals tends to be obsessed with controlling nature, merely for the sake of attempting to exercise that control (see Holling and Meffe 1996). Management should, instead, focus on discerning and solving problems that populations might cause. It is appropriate to be aware of how certain problems are associated with abundance, but the *central focus* should be on the problem caused by an animal population, not the number of animals in the population.

Fortunately, the misunderstandings about *K* that were used to justify culling elephants of Kruger from 1967 to 1994 were mitigated in the 1999 plan of managing the elephants of Kruger. That plan indicated that culling would no longer be triggered by elephant abundance exceeding some prejudged value. Instead, decisions would be based on what would seem to be elephant's impact on plant biodiversity (Dickson and Adams 2009).

Some other wildlife managers working in other cases have also been too focused on numbers for their own sake. A notorious example is a case in which managers in the US state of Wisconsin became needlessly bound by their own policy to have at least 350 wolves in the state, but no more than 350 wolves. This demand for 350 wolves occurred throughout two decades (2000–2020) during which abundance rose from 350 to about 1,000 wolves. The circumstance led to senseless killing of wolves and distracted attention from discerning and treating the problematic relationships between wolves and humans.

Overabundance of certain populations is a serious threat to many conservation goals. But one cannot just claim that a population is overabundant and prejudge that killing animals is the solution. Most problems involving so-called overabundant populations are a great deal more complex, and too often that complexity is not adequately treated.

Goals and Methods

In the elephant culling argument, premises P3 and P4 represent the ever-applicable notion that ethical actions require not only a worthwhile goal, but also acceptable methods for realizing that goal. The elephant culling case illustrates the varied considerations that arise when attending to that notion.

Concern for the harm caused by lethal methods used to control animal populations has motivated some to develop methods that are less harmful, such as sterilization and contraception. These can in fact be useful for some species in some situations. For example, there are promising signs that such methods can be used to reduce the abundance of gray squirrels in United Kingdom, where they are a non-native species (Rowlatt 2022). At present, however, these methods seem infeasible for limiting the abundance of elephants in a population as large as Kruger's. Challenges include the logistical difficulty associated with the large number of elephants that would be involved and concern that those methods could result in hormonally induced changes in behavior that would be detrimental to elephants. With future research, sterilization and contraception could become feasible for elephants, but that seems not to be the case today.

Culling in Kruger had involved killing about 550 elephants annually. In Chapter 5, you'll learn how to determine which individuals in a population contribute most to population growth. If managers had taken advantage of such information, they would likely have been able to limit abundance by killing 300 elephants annually (Whyte, van Aarde, and Pimm 1998; van Aarde, Whyte, and Pimm 1999). To do so, they would have focused on killing young female elephants.

What if killing 300 elephants per year really is the least harmful way to limit abundance? What if no further reduction in harm is feasible? Identifying the least harmful method is important, but it is not the same as judging whether that least harmful method causes more harm than can be justified.

Making that judgment requires developing and evaluating a separate argument, whose conclusion would be, for the elephant case, something like

"The benefits to be realized from killing x number of elephants annually outweigh the harm of that killing." Arguments aiming to adjudicate the competing values of conservation and the just treatment of individual animals are among the most difficult to assess. The need to assess such arguments is also increasingly common. Aside from that concern, do not overlook the central point of this section, which is that good goals do not necessarily justify harmful methods.

Ecosystem Health

Often judging the appropriateness of a conservation action that causes harm is aided by probing and better articulating the hoped-for benefits of the action. For the elephant culling case, the hoped-for benefit is restoration and maintenance of ecosystem health. The obvious question is, what precisely is meant by the phrase "ecosystem health"?

By one view, ecosystem health means letting ecological processes unfold naturally, where "natural" means without being influenced by humans. That idea can lead to the belief, for example, that elephants should be killed to protect the natural process of herbivory and its influence on plant communities. That is, they should be killed to prevent unnaturally high densities of elephants from browsing at an unnaturally high intensity.[5]

Those ideas prompt an appropriately contrarian question: If natural processes are so deeply valued, then why not allow for natural fluctuations in elephant abundance? That is, why not allow elephant abundance to fluctuate without the interference of humans killing elephants. On what grounds does one decide that one natural process (herbivory) is more important to protect than another (elephant population dynamics)?

The challenge to answering such questions fuels controversy elsewhere. One example concerns whether wolf predation should be conserved in Isle Royale National Park in Michigan (USA) on occasions when extinction of the wolf population seems imminent (Vucetich 2021). On the one hand, the park is an island, and extinction is a natural process on islands. On the other hand, predation is a natural process, and humans played an important role in its loss. The natural process that is judged to be more important can have an important influence in judging which conservation action should be taken.

Reflecting on questions about what's natural for the elephant case may lead to thinking that the preservation of "natural processes" is not the underlying motivation for culling elephants. Perhaps the motivation is narrower: If elephant abundance remains high in Kruger, then some plant species risk being extirpated from Kruger. If that is the concern, how does one weigh the harm of killing hundreds of elephants a year against the harm that elephants might be causing to those plants?

Addressing that question may be aided by recognizing that elephants and forests "naturally" coexisted in the past, before humans reduced and fragmented the geographic range of elephants (Fig. 2.1). That coexistence may have depended on dynamics occurring over very large landscapes and long periods of time. At some places and times, elephants may have been "naturally" abundant and had a significant impact on vegetation. Eventually, in the absence of food, the elephants could not survive, and their abundance "naturally" declined at that place and time, after which, the vegetation recovered. In the meantime, elephant abundance increased at some other place. That vast spatial-temporal dynamic may have been the natural process that was broken and in need of restoration.

The concern is that elephants have no blame in that loss. That loss is entirely on humans. That being the case, why should elephants pay the price for on-going loss caused by humans?

One might respond by observing that it's not possible for humans to reverse those losses, especially given that Africa's population of humans is expected to nearly double over the next 30 years. Yes, agreed. But that last question did not ask whether humans were able (or willing) to undo the damage. The question is, in essence, Is it wrong to make elephants pay for something that humans did?

My raising those questions is not an unspoken prejudgment about the answers to those questions. And tending to such questions is important for understanding what actions are best in this case about elephants, as well as myriad other cases in the conservation of animal populations.

To this point, argument analysis has led us to recognize three important issues that can be generalized and summarized as follows:

· Ecosystem health is a complicated and unresolved idea. That circumstance rises from ecosystem health perhaps seeming to be an entirely scientific idea. Yet, careful consideration reveals the idea to be a tight blend of science and values. That blending is what makes the idea difficult to work with. That difficulty is frustrating because ecosystem health would appear in many arguments as the motivation for a wide range of conservation actions. In Chapter 6, we'll see that answering the question "What is an endangered species?" is fraught with similar challenges.

· We need to understand conservation motivations—such as maintaining ecosystem health—with so much rigor because many conservation actions involve harm. In other words, actions designed to honor the value of conservation often conflict with other values. Being able to say exactly how and why some conservation action is of value is important for adjudicating such conflicts and evaluating premises such as P4′.

· An increasingly common conflict is between conservation and the fair treatment of nonhuman animals. The concern is that over-abundant populations are one of the broadest threats to conservation, and a common response to overabundant populations is killing individuals of the overabundant population. The concern is complicated because humans have not yet found broad agreement about what it means to treat nonhuman animals fairly in the context of conservation.

Scientific Uncertainty

Premise P2′ of the elephant argument is a scientific claim about which there is scientific uncertainty. One uncertainty is the extent to which changes in plant communities are attributable to drought as opposed to elephant herbivory. There is also uncertainty about how long it would take for plant communities to return to their former states after having been changed by elephants (or drought).

If P2′ is false, then the argument collapses and fails to support elephant culling. That statement is more important than may be obvious. To judge whether elephant culling is appropriate for some given reason, would seem—on first appearances—to depend entirely on details about ethical premises. But we can now see that support for an ethical conclusion can hinge greatly on a scientific premise. If P2′ were shown to be false, then there might, for example, be no need to consider whether the ethical premise of P4′ is appropriate, because the failure of one premise is enough to cause the complete failure of an argument.

Uncertainty about P2′ raises another concern. Should one assume that P2′ is false until proven true? If so, one would refrain from culling until P2′ is proven true. Or, should one presume that P2′ is true until proven otherwise? These questions pertain to what is called *burden of proof*.

A common response to this circumstance is to invoke what is known as the *precautionary principle*, which states that one should refrain from actions expected to result in irreparable harm. That's a good idea, but it does have a severe limitation. Specifically, advocates of elephant culling would likely claim that changes to the plant community are the irreparable harm to avoid. At the same time, those opposed to elephant culling would likely claim that killing elephants is the irreparable harm to avoid.

Decision-making in the presence of scientific uncertainty is also the subject of much research. Useful as that research is, it does not entirely resolve ethical concerns pertaining to the precautionary principle.

To go further on this subject is beyond the scope of this book. For now, it is enough to know that this is a major theme in conservation ethics and that argument analysis is useful for exposing where and how scientific uncertainty can influence a decision. In the online supplement, you'll find guidance to continue your learning on this subject.

Ethical Inquiry and Conservation Decision-Making

Science has long been recognized as deserving an important role in making conservation decisions. That appreciation is accompanied by recognition that science is an ongoing process with no definite end, but conservation decisions need to be made in a timely manner—sometimes before all the relevant science is available. Furthermore, delayed decisions are usually de facto decisions to maintain the current course of action (or inaction). These concerns have led to what is now called *adaptive management*, which is built on the solid idea that timely decisions should be made on the basis of science available at the time, and revised decisions can be made when and if new science suggests the need to do so. Science and decision-making are joined to one another in an on-going process with no definite end (left-hand grouping of Figure 4.5).

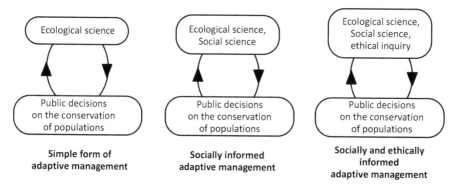

Figure 4.5 The evolution of different kinds of adaptive management. A long-appreciated idea is that ecological science and conservation decision-making should be engaged with each other in an on-going process (left panel). Appreciation has since grown to also account for insight from the social sciences (middle panel). A bright future for conservation decision-making would also include formal ethical inquiry (right panel). The upward-pointing arrows refer to evaluation of the consequences of a policy; the downward-pointing arrows refer to revision of policy based on the evaluation. Adapted from Vucetich and Nelson (2012).

Appreciation for adaptive management began in the 1980s. By the 1990s, there was increasing awareness that good decisions depend on both science and values. That led decision-makers to sometimes include social scientists as part of the decision-making process to provide, for example, scientific descriptions of the values that people hold and scientific descriptions of why people hold those values. This information is part of what is known as the *human dimensions of wildlife conservation*, and its inclusion in decision-making might be called *socially informed adaptive management* (middle grouping of Figure 4.5).

Only modest wisdom is required to know that what we value is not necessarily what we ought to value. Thus, it is important to understand the gap between what we value and what we ought to value.[6] Whereas, the social and psychological sciences are important means of describing our behaviors, ethics and its main tool, argument analysis, are important means for understanding how we ought to behave.

For those reasons, there is increasing interest in infusing conservation decision-making with insights from ethics (right-hand grouping in Figure 4.5). Because ethics and argument analysis are on-going processes (like science), their inclusion in the decision-making process is very much like the inclusion of ecological science and social science. Decisions should be informed by science and ethical inquiry, and decisions should be made in a timely manner. As new scientific or ethical insight becomes available, prior decisions can be revised.

From these considerations, two general points arise. First, the inclusion of ethical inquiry is a natural progression of increasingly informed decision-making. Second, the ongoing, indefinite nature of ethical inquiry is no more an obstacle for making timely decisions than is the ongoing, indefinite nature of science.

Some additional thoughts about ethical inquiry and argument analysis. You may, at this point, feel awash with concerns and unanswered questions about ethics and argument analysis. If so, that would be appropriate. Ethical inquiry is much like ecological science: learning a little leads a curious mind to deeper questions.

The comparison between ethical inquiry and science runs deep. Science is difficult to learn, which is why it takes four years to earn a bachelor's degree in science. Argument analysis, which is essentially logic, and the formal principles of ethics are also difficult to learn. You wouldn't expect to be good at science or have all your questions answered after taking one science class. The same expectation should apply to learning about ethical inquiry and argument analysis.

You might also anticipate that ethical inquiry and argument analysis are unlikely to definitively resolve, once and for all, any decision to which they are applied. That is correct. But definitive resolution is not the goal of any instance of argument analysis—not any more than a particular scientific research project is expected to be the final word on any matter of science.

Rather the value of ethical inquiry and argument analysis includes the clarity, precision, and transparency they offer for the reasons that motivate our actions. An intense interest in explaining our reasons is essential for maintaining our humanity.

Finally, I draw your attention to the online supplement for this chapter, which includes discussion questions and exercises to reinforce what you've learned from it, as well as guidance for further reading on the topics covered here.

RETURNING TO THE ELEPHANTS OF KRUGER

Our partial analysis of the elephant culling argument revealed important issues that may have been overlooked without argument analysis. We also discussed the importance of making decisions on the basis of assessments that are incomplete, provisional, and fallible. With that mind, you might wonder about the consequences of having decided to begin culling in 1967 and to stop culling in 1995.

You already know that elephant abundance rose far beyond what was said to be the population's carrying capacity (Fig. 4.3). In 2003, eight years after elephant culling stopped, managers closed about two-thirds of the watering holes that humans had been maintaining in the park. Yes, I forgot to mention that humans had for decades been providing water to wildlife in a xeric (dry) environment. That brings at least a twinge of sad irony, given that some of the reason to cull elephants may be aptly framed as an interest in letting natural processes dominate.

After 2003, elephants spent less time foraging near the closed watering holes, and the vegetation seems to have begun recovering. Elephants also began to spend more time near some rivers, where herbivory had a greater impact on vegetation. In those places, managers have been experimenting with nonlethal ways to deter elephants from spending so much time there. The methods include making loud noises, like gunshots, and setting up bee hives, which are a bother to elephants.

The overall impact of elephants on vegetation since culling stopped is far from obvious and still being studied. But management has shifted from trying to control abundance by killing to trying to manage the spatial distribution of elephants by nonlethal methods.

The past 20 years seem to suggest that P3 from the elephant culling argument—that the plan for killing elephants adequately limits harm to ecosystem health and does so with the least harm—was false. What's difficult to judge is this: Were thousands of elephants killed because the *science* was not good enough to have prevented managers from thinking it a good idea? Or were thousands of elephants killed because the harm of that killing was underappreciated to the point that managers were lax about developing nonlethal methods for managing the elephants of Kruger? The reason to consider such questions is less to judge past decisions and more to inform other ongoing cases in which decisions are made to kill (or to refrain from killing) in the name of conservation.

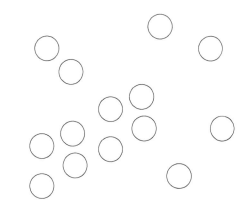

CHAPTER 5

Structured Populations

In the waters of Puget Sound, near Washington and British Columbia, lived an orca distinguishable to human researchers from other orcas in her pod by a distinctive white patch just behind her dorsal fin and a small nick with a protruding bit of flesh on the trailing edge of her dorsal fin. This orca is identified in record books as J2, but researchers also knew her as Granny. She was first observed in 1976, when researchers began studying this pod. Over many years of observation, she never reproduced. Eventually researchers deduced that she was postmenopausal and that the last time she gave birth was sometime before 1976.

Granny was also the leader of J-pod (Fig. 5.1). That leadership is no small role, given that orcas are also among the most social of all mammals. Relationships within and between pods are complex. Only elephants, chimpanzees, and humans have social lives as involved as orcas.

Over the years Granny had been seen with J-pod thousands of times. Then, after October 12, 2016, when she was last sighted, she disappeared. By December she was presumed dead, her body lost to the vast waters of the Pacific Ocean.[1]

That she lived to be perhaps 80 years old is impressive. But that she lived about half her life in a post-reproductive state is especially remarkable. Among the 5,400 species of female mammals, only humans, chimpanzees, and 4 species of cetaceans, including orcas, are known to experience a post-menopausal life.

Female orcas give birth about once every 5 years between the ages of about 10 and 40. Before age 10 they are sexually immature; by age 40, most

Figure 5.1 The orca known as J2, aka Granny, with members of the pod that she led. Credit: Center for Whale Research. Photo taken under NMFS Permit 21238/DFO SARA 388 of the Canadian Government.

female orcas are postmenopausal. It was recently discovered that orcas have higher survival rates and reproduce more successfully if they live in pods led by a postmenopausal female (Nattrass et al. 2019).

Orcas are a distinctive example of a common circumstance. The contributions that an individual makes to a population's dynamics depends on their age. Conversely, the dynamics of a population depend greatly on the ages of individuals within a population. In this chapter we're going to learn how both dynamics work.

Our overarching goal continues to be understanding how and why populations fluctuate in abundance over time. We pursued that goal in Chapters 2 and 3 for populations in which each individual in the population can be reasonably treated as identical to every other individual—identical, that is, in terms of their vital rates, specifically, individuals' chances of surviving and reproducing in the upcoming year. In other words, individuals are treated as identical with respect to their contributions to the population's per capita growth rate, *r*. To consider a population in that way is to treat it as *unstructured*. Doing so can lead to valuable insight, as it did in Chapters 2 and 3.

But the individuals of a population are typically not identical with respect to their vital rates, and additional insight can be gained by taking account of the differences among individuals. Taking account of such differences is to treat a population as *structured*.

CONCEPTUAL FOUNDATION

In Chapter 2, we learned that the annual per capita growth rate (r) can be decomposed into annual rates of recruitment and survival. Those vital rates tend to vary as an individual ages. Juvenile and senescent individuals tend to have lower rates of recruitment and survival than prime-aged individuals (Fig. 5.2). Consequently, r depends on the relative abundance

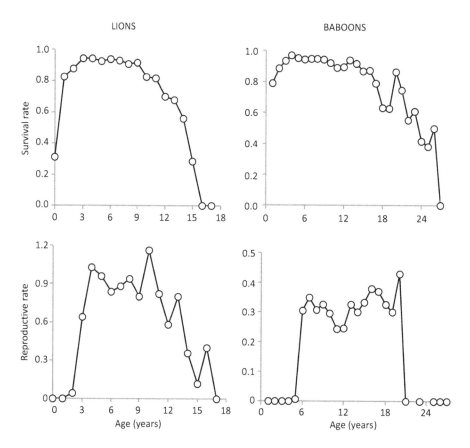

Figure 5.2   Age-specific vital rates for lions of Ngorongoro Crater, Tanzania, and female baboons of Amboseli, Kenya. The patterns are broadly representative of many longer-lived animals, for which survival and reproduction is greatest for prime-aged individuals. Adapted from Packer, Tatar, and Collins (1998) and Altmann and Alberts (2003).

of prime-aged individuals within a population. If two populations with the same abundance and the same environmental conditions had different relative frequencies of prime-aged individuals, then the population for which prime-aged individuals were more frequent would very likely have a higher r.

I should clarify a few pieces of new vocabulary from the previous paragraph—just to be sure. Survival rate (s) is the mathematical complement of what I'd previously called the mortality or death rate (d). In other words, $s = 1 - d$. It will be obvious within a few pages why it is now useful to speak in terms of s rather than d. Senescence is a biological process by which one's body and bodily functions decline with age. Senescent individuals are those who have senesced. Prime-aged individuals are older than juveniles and younger than senescent individuals. Now that we've confirmed the vocabulary and because that previous paragraph is so important, I recommend giving it a quick re-read. The relative abundance within a population of different-aged individuals (such as juveniles, prime-aged, senescent) is a way of describing a population's *age structure*. Another way to describe age structure is, for example, by the average age of individuals within a population. Regardless of how age structure is measured, it fluctuates over time for many populations (Fig. 5.3, panel A). These fluctuations in age structure are associated with fluctuations in per capita growth rate (Fig. 5.3, panel B).

Why does age structure fluctuate over time? An important answer is that different-aged individuals are sometimes impacted differently by exogenous influences, such as predation and unfavorable weather. More specifically, predation and unfavorable weather sometimes impact juveniles and senescent individuals more than they do prime-aged individuals. If a population is exposed to a string of years with high predation or unfavorable weather, the result can be a reduction in the relative abundance of juveniles and senescent individuals (see Wilmers, Post, and Hastings 2007). Hunting can also alter the age structure of a population.

Let's recap. r depends on the vital rates of individuals within a population. Because these vital rates vary with an individual's age, r depends on a population's age structure, which often fluctuates over time. These fluctuations are often due to different-aged individuals responding differently to changing environmental conditions.

Before the end of this chapter, we'll see that structured population dynamics—of which age-dependent population dynamics are one kind—are interesting, complex, and worth understanding for the sake of better conservation.

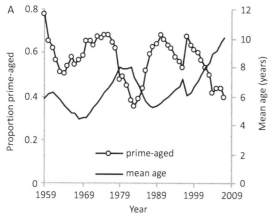

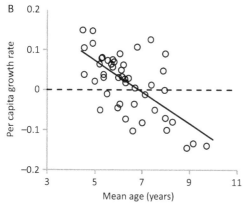

Figure 5.3 Fluctuation in the age structure of the moose population in Isle Royale National Park over a five-decade period (A). Primed-aged individuals are 2–9 years of age; for context, the oldest moose on Isle Royale lived to be about 19 years of age. Per capita growth rate tends to be much lower when the population's mean age is higher (B). A higher mean age is indicative that the population has many senescent individuals, who have lower rates of survival and reproduction. The regression line in B accounts for 43% of the variation in per capita growth rate. Adapted from Hoy et al. (2020).

MATHEMATICAL FRAMEWORK

A precise account of structured dynamics requires some math. To discover this math, we need a kind of translation process between the physical world and the world of math. For structured population dynamics, this translation begins with life-cycle diagrams.

Life-Cycle Diagrams

We'll find our way to the necessary math by first describing a population's structure with a life-cycle diagram. These diagrams are comprised of nodes that represent different kinds of individuals within a population. The upper diagram in Figure 5.4 is a life-cycle diagram for an *age-structured* population, for which individuals' maximum life span is five years.[2] Each node represents individuals of a different age. The numbers below each solid arrow represent the probability that an individual will survive from one year to the next and in the process become one year older. Notice that those probabilities of survival change with an individual's age. That there is no solid arrow leaving the last node indicates that all of those individuals die

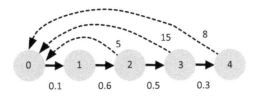

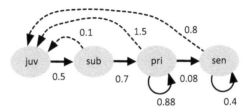

Figure 5.4 Two life-cycle diagrams. The upper diagram depicts an age-structured population for a species whose maximum life span is five years. The lower diagram represents a stage-structured population, where the stages are different age classes—juvenile (juv), sub-adult (sub), prime-aged (pri), and senescent (sen). See main text for further explanation.

by the end of that year (due to old age). The dashed arrows pointing to the first node, represent the mean number of recruits produced by individuals of each age in the upcoming year.* Notice that the contributions to reproduction are also age-dependent. This diagram further indicates that individuals reproduce for the first time when they are two years old.

The lower diagram of Figure 5.4 is an example of a life cycle for a *stage-structured population*, for which individuals are well-described as belonging to one of four age classes: juveniles (juv), sub-adults (sub), prime-aged (pri), and senescent (sen). A key difference between age structure and stage structure is that stage-structured populations allow for one or more stages, where an individual may remain in the given stage for more than a year. For the example given in Figure 5.4, individuals can be a juvenile for only a year and a sub-adult for only a year. But individuals may spend multiple years in the prime-aged stage and multiple years in the senescent stage. The possibility of remaining in a stage for multiple years is represented by the solid arrows that originate and terminate with the same node.

Focus for a moment on the solid arrows that leave the node for prime-aged individuals. The arrow that begins and ends with the prime-aged node is associated with the value 0.88, which means that each year 88% of those individuals will, on average, remain in that stage for the upcoming year. The arrow that goes from prime-aged to senescent is associated

---

* Recruits are offspring that are born or hatched and survive to the end of their first year.

with the value 0.08, which means that each year 8% of the primed-aged individuals will, on average, transition into the senescent age class. In this case, the annual survival rate of prime-aged individuals is the sum of values associated with solid arrows leaving that node, which in this case is $0.88 + 0.08 = 0.96$. The fact that 0.88 is so much larger than 0.08 is indicative that individuals can spend a number of years in this age class.

Now focus on the senescent node in the lower diagram of Figure 5.4. The only solid arrow emanating from that node terminates at that same node, and the value associated with that arrow is 0.4. This means that 40% of the individuals in that stage will survive the year and spend the upcoming year in that same stage. The remaining individuals die. From that last stage, there is no other stage to transition to.

The magnitude of the values associated with arrows circling back onto the same life stage may be interpreted as just described. But those values also reflect the duration of the stage. Larger values are indicative of a life stage for which an individual can spend more years. Knowing the details of that relationship are not necessary for gaining the key insights of this chapter. Nevertheless, for those who are interested I offer some details in the online supplement.

After considering these two kinds of diagrams, note this important, but subtle distinction:

· In the age-structured life cycle, the annual survival rate of an individual is represented by the numbers associated with the solid arrows.
· In the stage-structured life cycle, the annual survival rate of an individual for any given stage is the sum of numbers associated with solid arrows emanating from that stage.

Furthermore, the values associated with the stage-structured life cycle in Figure 5.4 are broadly representative of an ungulate with a life span of approximately 20 years. It is possible to represent such creatures with an age-structured life cycle, but doing so requires a large number of nodes and arrows. The advantage of a stage-structured life cycle is its ability to describe a long-lived species with relatively few nodes and arrows.

Projection Matrices

You are familiar with performing various arithmetic operations such as addition and multiplication on *numbers*. There is also a branch of mathematics known as *matrix algebra* that involves performing similar operations on *sets of numbers all at once*. To analyze structured populations, we need to use some matrix algebra.

If you like math only so far as you can see how it helps us understand population biology, that's okay. But even so, you might need a pep talk at this moment. We need to work through about nine pages of unavoidable math before we can see how useful it is. None of the math is difficult, but it does take a little patience (nine pages' worth). If you are patient with the math, you will not be disappointed. Some very fine insights about real populations are coming.

Structured population dynamics are connected to matrix algebra because a life-cycle diagram can be represented as a *projection matrix*. A matrix is a collection of numbers arranged in rows and columns. A projection matrix represents how individuals transition through different stages of a life cycle, contribute to the creation of new individuals as they transition through life, and then eventually die. We'll soon be using this matrix as a tool for projecting a population's abundance from one year to the next.

Next, we need to learn a simple skill—namely, how to translate a life-cycle diagram into a projection matrix. The first thing to know is that a projection matrix always has the same number of rows and columns. It is square. Furthermore, the number of rows and columns is equal to the number of nodes in the life cycle. So, the species with a five-year life span would have a matrix with five rows and columns. We say it is a 5×5 matrix. The species described as having 4 life stages would have a 4×4 matrix.

The two life cycles in Figure 5.4 are presented as projection matrices in Figure 5.5. Before I explain the details, make a quick comparison of Figures 5.4 and 5.5. You may notice some patterns. For example, note that

$$A = \begin{bmatrix} 0 & 0 & 5 & 15 & 8 \\ 0.1 & 0 & 0 & 0 & 0 \\ 0 & 0.6 & 0 & 0 & 0 \\ 0 & 0 & 0.5 & 0 & 0 \\ 0 & 0 & 0 & 0.3 & 0 \end{bmatrix}$$

$$A = \begin{bmatrix} 0 & 0.1 & 1.5 & 0.8 \\ 0.5 & 0 & 0 & 0 \\ 0 & 0.7 & 0.88 & 0 \\ 0 & 0 & 0.08 & 0.4 \end{bmatrix}$$

Figure 5.5 These projection matrices correspond to the life-cycle diagrams of Figure 5.4. The arrangement of numbers in a projection matrix allows for predicting the population's dynamics. See main text for details.

0.7 in row 3 and column 2 of the lower matrix corresponds to transitioning from stage 2 (sub-adult) to stage 3 (prime-aged).

Each number in the diagram has a special place in the matrix. The remaining places in the matrix are zeroes. More specifically, numbers involving reproduction appear in the top row. Numbers representing individuals that stay in the same life stage for two years in a row appear along the main diagonal of the matrix. And numbers representing the graduation from one stage to the next appear along what is referred to as the *subdiagonal* or *off-diagonal*, just below the main diagonal. These patterns are most quickly apprehended if the numbers are color-coded. I recommend using a set of yellow (for elements on the top row), blue (for the main diagonal) and green (for the sub-diagonal) highlighters to show yourself how the numbers in Figures 5.4 and 5.5 correspond to each other.

A more precise way to understand the projection matrix is aided by some notation and vocabulary. First, let the projection matrix be represented by the symbol A. A number in a matrix is called an element and denoted $a_{i,j}$, which refers to the element in the ith row and the jth column of the matrix.[3] Next, distinguish the first row of the matrix from the remaining rows:

- Elements in the first row involve contributions to reproduction made by individuals from each age (or stage) of the life cycle. For example, element $a_{1,3}$ in the 5×5 matrix of Figure 5.5 is 5, meaning that each two-year-old contributes, on average, 5 recruits to the population each year.
- Elements in the subsequent rows describe how individuals survive, age, and die. More precisely element $a_{i,j}$ represents the probability that individuals of the jth age (or stage) transition to the ith age (or stage) over the course of the next year. For example, in the 4×4 matrix of Figure 5.5, element $a_{4,3}$ describes the expected proportion of individuals in the third stage (prime-aged) that will survive and transition into the fourth stage (senescent) over the upcoming year. Element $a_{3,3}$ describes the proportion of individuals in the third stage (prime-aged) that will survive and remain in the same life history stage over the upcoming year. In more colloquial terms, elements *not* in the top row describe survival and transitioning *from* the age (or stage) indicated by the element's column *to* the age (or stage) indicated by the element's row.

Caveat. For creatures that reproduce annually, the elements of a projection matrix are derived from estimates of annual survival rate. The details for doing so depend on *when* during the year the population is assessed for survival. What I've described is appropriate for cases in which survival is assessed each year, at a time of year just *before* organisms are born, which means the first age class (labeled 0 in Figure 5.4) is the number of animals that were born *and* survived to the cusp of their first birthday.

In some cases, however, survival is assessed each year just *after* organisms are born. In these cases, elements in the matrix are determined in a slightly different fashion. While those details are covered in the supplementary materials, the important messages of this chapter are readily accessible without minding them.

Population Vectors

While a projection matrix defines how individuals survive and transition through ages (or stages) of a life cycle, a *column vector* represents the number of individuals in a population belonging to each age or stage at some point in time. A column vector is a special kind of matrix, distinguished as being a single column of numbers. The number of rows in this column vector is equal to the number of ages or stages in the life cycle. When a column vector is used in this way is it usually denoted as N.

A note about notation. Matrices and vectors are often denoted by bold, italicized capital letters, such as N and A. The elements within a matrix or vector are often denoted by regular italicized lowercase letters with subscripts. For example, n_3 refers to the third element of N and $a_{i,j}$ refers generically to the element of A in the ith row and jth column.

With such notation, subscripts do not refer to time. If one wants to refer to time, as in abundance at time t or $t+1$, then one writes $N(t)$ or $N(t+1)$.

The notation takes a little getting used to, but like reading Shakespeare, it feels pretty natural after a little practice. And the notation does serve an important purpose, which is, to communicate a verbose idea, such as *the number of three-year-olds in next year's population*, with just a few keystrokes: $n_3(t+1)$.

While the details of notation vary among authors, what is used here is representative.

Matrix Multiplication

If one multiplies the projection matrix, A, times the population vector, $N(t)$, the result is a prediction for next year's population vector, $N(t+1)$. The procedure for matrix multiplication is depicted in Figure 5.6.

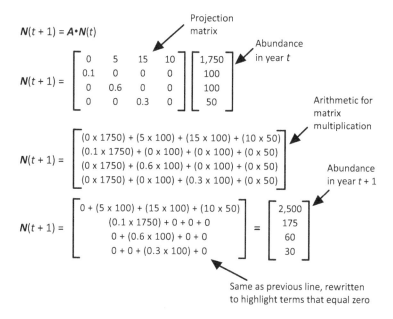

Figure 5.6 Matrix multiplication for predicting next year's abundance, N(t + 1), from the projection matrix, A, and the population vector that describes this year's abundance, N(t). See text for details.

In Figure 5.6, notice that A is a 4×4 matrix and $N(t)$ is a 4×1 vector, meaning it has 4 rows and 1 column. Our goal is to compute a number for each of the four elements in $N(t + 1)$.

Also, notice the arrangement of non-zero elements in A. There are non-zero elements along the sub-diagonal, but elements of the diagonal are all zeros. This arrangement of non-zero elements is indicative of an age-structured population (not stage-structured) in which the maximum life span is four years and two-year-old individuals are the youngest that can reproduce.

The first step in matrix multiplication is to calculate the element $n_1(t + 1)$, which is the number of newly recruited individuals in the upcoming year. This number is the sum of recruits contributed by individuals of each age. In other words, calculate four products (using regular multiplication) and add them together (using regular addition). Symbolically it looks like this:

$$n_1(t + 1) = (a_{1,1} \times n_1) + (a_{1,2} \times n_2) + (a_{1,3} \times n_3) + (a_{1,4} \times n_4).$$

In the preceding equation and those that follow, I suppress the use of t on the right side of the equation so that the equation is easier to read. But you can know, for this passage of text, that time is $t + 1$ on the left side of the equations and t for the right side. When that calculation is performed in Figure 5.6 the result is 2,500 newly recruited individuals.

You might be thinking, my goodness, this is going to be a lot of multiplication and addition, just to predict next year's abundance. Yes, it is. And you'll want to do it just a few times "by hand" so that you understand how it works. Afterward you'll learn how to perform these calculations more efficiently in Excel. But for now, back to the arithmetic.

If this text feels confusing, don't be concerned. It'll make sense when you use this text, line-by-line, to guide you in performing a calculation like this on your own, and the chance to do so is coming soon.

The next element of $N(t+1)$ to calculate is $n_2(t+1)$ which is the number of this year's recruits that survive and transition to the next age over the course of the upcoming year. The element, $n_2(t+1)$, is calculated similarly to $n_1(t+1)$, except this time we'll be working with the second row of the projection matrix (whereas we previously worked with the first row). More precisely, calculate the four products and add them together:

$$n_2(t+1) = (a_{2,1} \times n_1) + (a_{2,2} \times n_2) + (a_{2,3} \times n_3) + (a_{2,4} \times n_4).$$

Notice that most of these products in Figure 5.6 are 0:

$$n_2(t+1) = (0.1 \times 1750) + (0) + (0) + (0)$$

Most of the products are 0 because element $n_2(t+1)$ is the number of two-year-old individuals at time $t+1$ and that number depends only on the number of recruits at time t. In other words, the previous equation indicates that of the 1,750 recruits in the population at time t, 10% of them will survive the upcoming year and become two-year-old individuals.

In broader terms, and speaking of age-structured models (not stage-structured), the number of individuals of a certain age in the upcoming year depends only on the number of individuals of the previous age from the current year. For example, the number of three-year-old individuals alive at time t has no bearing on how many two-year-old individuals there will be at time $t+1$.[4]

The last two elements, $n_3(t+1)$ and $n_4(t+1)$, are calculated similarly, except that calculating $n_3(t+1)$ involves the third row of the projection matrix and calculating $n_4(t+1)$ involves the fourth row. For each case, you sum four products, most of which are zero:

$$n_3(t+1) = (a_{3,1} \times n_1) + (a_{3,2} \times n_2) + (a_{3,3} \times n_3) + (a_{3,4} \times n_4),$$
$$n_4(t+1) = (a_{4,1} \times n_1) + (a_{4,2} \times n_2) + (a_{4,3} \times n_3) + (a_{4,4} \times n_4).$$

That's it. Now, I recommend that you practice with the simple example shown in Figure 5.7.

$$N(t + 1) = A \cdot N(t) = \begin{bmatrix} 0 & 3 & 6 \\ 0.2 & 0 & 0 \\ 0 & 0.6 & 0 \end{bmatrix} \begin{bmatrix} 500 \\ 300 \\ 150 \end{bmatrix} = \begin{bmatrix} n_1 \\ n_2 \\ n_3 \end{bmatrix}$$

[Figure 5.7]

Your goal is to calculate values for n_1, n_2, and n_3. The answers are given at the end of this chapter. To further show yourself how much you've learned, you could also draw the life-cycle diagram for an organism with a projection matrix like that shown in Figure 5.7.

The preceding examples are for age-structured populations. The multiplication procedure for stage-structured dynamics is the same. The only difference is that there tends to be fewer products that are zero. This is because, for example, if senescent individuals can be senescent for several years, then next year's abundance of senescent individuals ($n_{sen}(t+1)$) depends on this year's abundance of senescent individuals ($n_{sen}(t)$) as well as this year's abundance of prime-aged individuals ($n_{pri}(t)$).

Finally, if you go to YouTube and search "multiply matrix times vector", you'll find worthwhile videos that reinforce what I've described here.

This idea will make more sense after you make this kind of calculation on your own, and that opportunity is provided in the online supplement.

Lambda (λ)

Populations with projection matrices like those we've been discussing are density-independent. Abundance either increases exponentially to infinity or decreases asymptotically to zero. In other words, these populations do not have a carrying capacity. Mathematicians would say these populations do not have an *equilibrium abundance*.[5] It is possible to create a projection matrix that is density-dependent, but we won't get into that here, in part because plenty of insight is to be had by studying structured populations that are density-independent.[6]

The rate at which abundance increases (or decreases) depends on the values of the elements in A. The general idea is intuitive. For example, increase the value of elements that reflect survival in A, and the population's per capita growth rate will increase. A more precise quantification of population growth rate in a structured population is, not surprisingly, more complex.

In particular, mathematicians learned long ago (before desktop computing) how to perform another kind of operation on a matrix, the result of which is called a *dominant eigenvalue*, often denoted by λ (lambda). While this value has special meaning to mathematicians, it also happens to be the population's *finite rate of increase* when the population is at its equilibrium.[7] More specifically, λ is how many times larger the population is next year as compared with this year, and it can be computed for structured and unstructured populations as

$$\lambda_t = N_{t+1}/N_t.$$

For example, if $N_1 = 100$ and $N_2 = 112$, then $\lambda = 1.12$, which means the population grew to be 1.12 times larger over the course of a year. In other words, the population is 12% larger.

Another example is useful. If a population's abundance is 80 and the declines to 60 over the course of a year, then $\lambda = 60/80 = 0.75$. In this case, λ indicates that the population is 75% of the size it had been in the previous year.

Referring to λ as the *finite rate of increase* distinguishes it from r, the *per capita growth rate*. These two measures of population growth, λ and r, are related to each other by these two equivalent equations:

$$\lambda = e^r; \tag{5.1}$$
$$r = \ln(\lambda). \tag{5.2}$$

You already know that abundance increases when r is positive and decreases when r is negative. For λ, the boundary between increasing and decreasing is 1. Abundance declines when $\lambda < 1$ and increases when $\lambda > 1$.

Note that because λ is the dominant eigenvalue of A, population biologists have a tendency to describe structured population dynamics in terms of λ. However, structured and unstructured populations can be described in terms of either r or λ. Moreover, it is possible to derive Equation 2.4 in terms of λ.

Equilibration

Envision building a projection matrix, A, based on knowledge of individuals' age- or stage-specific rates of survival and contributions to recruitment. Further suppose that A describes a population poised to increase. Now suppose you have an initial estimate for the population's abundance and age structure—that is, an initial estimate of N at time $t=0$. Finally, envision, projecting N into the future for, say, each of the next 10 years. When you do this, you'll observe several changes:

· Total abundance increases over time.
· The finite rate of increase (λ_t) is likely to change each year, but the changes get smaller each year as λ_t approaches its equilibrium value.
· The relative abundance of individuals belonging to different ages (or stages) changes each year, but the changes get smaller each year as the population's age (or stage) structure approaches its equilibrium.

The equilibrium age structure is sometimes called the *stable age structure*. It can be estimated by projecting a population's abundance over a number of years. The equilibrium is often reached within 10 to 20 years. These calculations became easy to make as desktop computing became widely available. Before that time, however, mathematicians had discovered how to perform another kind of operation on a matrix, the result of which is called a *dominant eigenvector*, which happens to be equal to the population's equilibrium age structure.

To recap, the value of λ_t depends on the elements in A and $N(t)$. Changes to any of the elements in A or $N(t)$ can affect the population's growth. While these populations do not have an equilibrium abundance (like K in Equation 3.3), they do have equilibria for λ and the population's age (or stage) structure. These equilibrium values, it turns out, depend entirely on the elements within A.

Sensitivities and Elasticities

As indicated, changing the elements of A can lead to changes in λ. Furthermore, changing some elements can result in large changes to λ, while changing other elements can lead to trivial changes in λ. The influence an element has on λ (be it great or small) is quantified by *sensitivities* and *elasticities*.

Calculating the sensitivity for an element of A with respect to λ requires applying the ideas of calculus to matrices. If you have had exposure to some calculus, you may find it useful to see that the equation for the sensitivity of element $a_{i,j}$ is

$$s_{i,j} = \partial \lambda / \partial a_{i,j}. \tag{5.3}$$

All that we need to know is what this equation means in English. In particular, Equation 5.3 answers the question, *How much does the population's growth rate change when the element $a_{i,j}$ changes by a small amount?*

To repeat, each element of *A* has its own sensitivity. That is, small changes in some elements can lead to big changes in the population's growth rate, but the same changes in other elements might lead to very little change in the population's growth rate.

Sensitivities have an important limitation. That is, direct comparisons of sensitivities for different elements of *A* are not always so informative, because the different elements sometimes represent different units and scales. For example, some elements are survival rates, such as 0.85/year. But other elements are contributions to recruitment, such as 3.5 recruits per year. Those differences can complicate the interpretation of sensitivities.

That limitation can be compensated for by comparing the elasticities of elements in *A*. The elasticity of an element is computed as

$$e_{i,j} = s_{i,j} \, (a_{i,j} / \lambda).$$

Elasticities answer the question, *What is the proportional change in population growth given a proportional change in $a_{i,j}$?*

An example will help. Suppose we are considering a population whose dynamics are described by a 5×5 age-structured matrix. The kinds of questions we can address with elasticities are as follows: If there is a 5% increase in reproduction for two-year-olds ($a_{1,3}$), by what percentage would population growth rate increase? Or, if there is a 5% increase in survival of one-year-olds ($a_{3,2}$), then by what percentage would λ increase?[8]

For emphasis, the most important thing to know about sensitivities and elasticities is the general concept represented by those italicized questions. You may already intuit how these ideas can be of value for better understanding, for example:

· the impact of a harvest on populations where individuals of one life stage are more likely to be harvested than those at other life stages.
· whether conservation strategies focused on improving conditions for individuals of some life stages may be more effective for

reducing extinction risk than improving conditions for those at other life stages.

If you cannot quite see how sensitivities and elasticities can be useful to such cases, no worries. Illustrative examples are coming next.

CASE EXAMPLES

Finally! We did it. We're done with what is arguably the most abstract portion of this entire book. Now we can think straightforwardly about our feathered, furred, and finned brethren. What comes next are some cases that make use of the mathy ideas presented above to gain insight about real animal populations.

Case 1: Comparing Structured Dynamics among Species of Birds

The great tit (*Parus major*) is a European bird much like the chickadee of North America (Fig. 5.8, left side).[9] Great tits are sexually mature after one year. A pair of adults fledge, on average, 3.2 offspring per year. Each fledgling has about a 20% chance of surviving to become a reproductive adult, and each adult has about a 44% chance of surviving the upcoming year. Typical of many small passerines, most great tits live no more than just a few years.

Northern fulmars (*Fulmarus glacialis*) are large oceanic birds of the North Atlantic, and they experience a very different kind of life (Fig. 5.8, right side). They take about eight years to reach sexual maturity. Females

Figure 5.8 Two species of birds with very different life histories: the short-lived great tit (left) and the long-lived northern fulmar (right). Credit: © Francis C. Franklin/ CC-BY-SA-3.0. https://en.wikipedia.org/wiki/File:Great_tit _side-on.jpg; https://upload .wikimedia.org/wikipedia/commons/4/43/Northern-Fulmar2_cropped.jpg.

Table 5.1 Vital rates for two bird species with different life histories. Recruitment is equal to the number of female offspring produced and surviving to the end of the nestling period per adult female per season. Maturity is the mean age (in years) of first reproduction.

Species	Recruitment	Maturity	Juvenile survival	Adult survival
Great tit	3.2	1 year	0.18	0.44
Northern fulmar	0.18	8 years	0.88	0.94

Figure 5.9 A life-cycle diagram and projection matrix that is generic enough to describe short-lived birds such as great tits and long-lived birds such as fulmars.

lay only a single egg per clutch, and both parents take turns incubating their egg for nearly two months. For every five mated pairs of northern fulmar, only about one successfully fledges a chick in a typical year. The annual survival rate after fledging is in the neighborhood of 0.9. Many northern fulmars live for a couple of decades.

These two species are compared more precisely in Table 5.1. For context, data for the great tit are derived from field observations like those described for the black-throated blue warbler (Chapter 3). Comparable data for large sea birds are derived by banding birds at nesting colonies and observing their nesting success and the proportion of banded birds that return to the colony from one year to the next.

As different as these two species are, they are both adequately represented by a life-cycle diagram comprised of just two life stages (juveniles and adults) and a 2×2 projection matrix (Fig. 5.9). The value of "forcing" such different species to match the same life cycle is to facilitate direct comparisons.

Simple as the projection matrix of Figure 5.9 may be, there is value in detailing its elements. First, the easy-to-describe elements appear in the second column of the matrix:

· Element $a_{1,2}$ is equal to recruitment.
· Element $a_{2,2}$ is equal to annual adult survival.

Both values are reported directly in Table 5.1. Elements in the first column of A are trickier:

- Element $a_{1,1}$ takes account of both the annual survival rate of juveniles and the duration (in years) of the juvenile life stage. Specifically, the value of $a_{1,1}$ is equal to $(1 - 1/T)s_{juv}$, where s_{juv} is juvenile survival and T is the number of years that individuals spend in the juvenile stage.
- Element $a_{2,1}$ is the proportion of juveniles that transition each year into the breeding adult stage. The element is equal to s_{juv}/T.

To derive the formulae for $a_{1,1}$ and $a_{2,1}$ is beyond the scope of this book. But it is important to know that in stage-structured matrices, the elements are typically combinations of vital rates or functions of vital rates.

With those details, you can determine on your own that the projection matrices for these two species are as shown in Figure 5.10.

$$\text{Great Tit} \qquad \begin{matrix} \text{Northern} \\ \text{Fulmar} \end{matrix}$$

$$A = \begin{bmatrix} 0 & 3.2 \\ 0.18 & 0.44 \end{bmatrix} \qquad A = \begin{bmatrix} 0.77 & 0.18 \\ 0.11 & 0.94 \end{bmatrix}$$

[Figure 5.10]

Notice that $a_{1,1}$ is much smaller for great tits than for fulmars (0 versus 0.77), as you would expect given differences in these species' life histories. Similarly, notice that $a_{2,2}$ is much smaller for great tits than for fulmars (0.44 versus 0.94), again as you would expect.

The online supplement includes a practice problem that guides you through the use of Excel to perform a simple kind of sensitivity analysis on those two matrices. I'll outline how to conduct such an analysis here. The first step is to set up a spread sheet like that shown in Figure 5.11.

You can use the spreadsheet to assess how changes in a vital rate result in changes to λ for great tits. For example, you can increase recruitment (cell A3) by 5%, from 3.2 to 3.36. and observe the change in λ (cell E20). Then reset recruitment to its original value and change adult survival (cell D3) by 5% (from 0.44 to 0.462) and again observe its impact on λ. Then you can perform similar calculations to see the effect of reducing those vital rates by 5%.

The last steps are to perform that analysis for northern fulmars and to graph the results in a way that highlights how the species differ with respect to the vital rates' influence on λ. That graph would look like Figure 5.12.

Two population ecologists from Norway conducted analyses similar to that represented by Figure 5.12. They did so for 49 species of birds for which they could find published data like that in Table 5.1. The result of that analy-

	A	B	C	D	E
1	**GREAT TIT**				
2	fecundity	age at maturity	juvenile survival	adult survival	
3	3.2	1	0.18	0.44	
4					
5	TRANSITION MATRIX, A		juvenile	breeding adult	
6		juvenile	0	3.2	
7		breeding adult	0.18	0.44	
8					
9			juvenile	breeding adult	
10	initial age structure	0.5	0.5		
11					
12			ABUNDANCES		
13		year	juveniles	breeding adults	total	lambda
14	0	50	50	100	1.91
15	1	160.0	31.0	191.0	0.7416
16	2	99.2	42.4	141.6	1.2167
17	3	135.8	36.5	172.3	0.9134
18	4	116.9	40.5	157.4	1.0706
19	5	129.7	38.9	168.5	0.978
20	6	124.4	40.4	164.8	1.029

Figure 5.11 The set-up for a spreadsheet to analyze structured dynamics in a population of great tits. Row 3 contains input values, taken directly from Table 5.1. The numbers in rows 6 and 7 are calculated from values in row 3 (see text for details). Cells B10 and D14 are initial conditions to set. Other values are calculated. For details, see the online supplement.

sis leads to one of the most important insights from studying structured population dynamics:

- Generally, vital rates can differ greatly with respect to their influence on a population's overall growth rate. Increasing survival by, say 5%, does not necessarily impact growth rate as much as changing the recruitment rate by 5%.

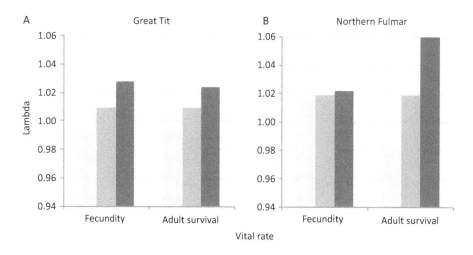

Figure 5.12 The set-up in Figure 5.11 can be used to assess how lambda is affected by small changes in each vital rate. The three bars rising above each vital rate (on the *x* axis) show how lambda increases with small increases in the vital rate. While *sensitivity* has a particular meaning in matrix algebra, one can also refer to graphs like these as representing a *sensitivity analysis*.

- More specifically, changes in adult survival for a long-lived creature tend to have a larger impact on λ than do changes in recruitment.
- For short-lived creatures, changes in recruitment tend to be similarly important or more important than changes in adult survival.

Case 2: Laysan Albatross

Wisdom was banded in 1956 on a tropical atoll about 1,300 miles northwest of Honolulu, Hawaii.[10] She was incubating an egg at the time, which means she was at least five years old. She was also observed on the same beach incubating another egg in February 2021.[11] Those observations mean that she is at least 70 years old and that she is the oldest known bird living in the wild. Over the decades, Wisdom has hatched about three dozen chicks. On her 6-foot plus wingspan, she's flown more than 3 million miles around the north Pacific—the equivalent of flying around the equator 120 times. She's made her life by searching for and eating cuttlefish, squid, the eggs of flying fish—almost anything floating near the ocean's surface. She even survived the Tōhoku tsunami that killed a couple thousand of her colony mates in 2011. Wisdom is a Laysan albatross.[12]

The world's population of Laysan albatross forages throughout the North Pacific. But each albatross nests in 1 of just 16 colonies, most of which are located on islands northwest of Hawaii. Forty-five percent of all Laysan albatross nest on the Midway Atoll where Wisdom has been nesting.

The world's population of Laysan albatross is also declining. A widely appreciated threat to the normally long-lived Laysan albatross is that too many are killed as adults when they get caught in fishing nets. The best available information suggests that between 1.9% and 5.0% of all adults drown each year.

The albatrosses of Midway suffer an additional threat. The main breeding site on the Midway atoll is Sand Island, which is also the site of a military base that was abandoned by the United States Navy in 1997. Since that time, the hundred or so buildings of that base have been deteriorating and shedding flakes of lead paint onto the nearby soil.

As albatross chicks become ambulatory, they also begin to exhibit inquisitive behavior, which unfortunately includes ingesting fragments of lead paint. Blood tests indicate that many chicks suffer severe lead poisoning. Many are so poisoned that they are never able to fly. They die at the end of the fledging season when their parents stop feeding them or sooner when other complications of lead poisoning set in, such as encephalopathy or renal failure. Taking account of how many chicks spend time near these buildings (about 10,000) and the estimated number of chicks on Sand island (146,000), one can figure that about 7% of chicks die from lead poisoning. Those deaths are in addition to deaths from natural causes.

The Midway breeding colony includes about 1.5 million Laysan albatrosses—representing about 60% of the world's population—and has been declining in abundance. Between 1991 and 2007, the colony's annual λ was estimated to be 0.995. That decline represents the situation before lead-induced chick mortality began affecting the population's abundance. Because albatrosses live so long, the impact of lead-induced chick mortality on the number of breeding adults will not be readily appreciated until long after the poisoning begins—perhaps 10 to 30 years after.

To better understand population dynamics of this breeding colony, a team of researchers from the United States built a 9×9 projection matrix based on the life cycle of the Laysan albatross (Fig. 5.13). They estimated the value of elements within the matrix from banding data collected throughout the second half of the twentieth century from birds like Wisdom. To understand how the colony would fare under different scenarios of lead-poisoned chicks and drowned adults, the researchers performed sensitivity analysis on two elements of the matrix ($a_{9,9}$, which represents adult survival, and $a_{1,9}$, which represents the fledging rate of new chicks).

The analysis indicates that lead-induced chick mortality causes λ to drop from 0.995 to 0.991. That decline in λ corresponds to population abundance declining by 36% over 50 years, rather than the 22% decline that would

$$A = \begin{array}{ccccccccc} \text{egg} & \text{1 yr old} & \text{2 yr old} & \text{3 yr old} & \text{4 yr old} & \text{5 yr old} & \text{6 yr old} & \text{7 yr old} & \text{adult} \\ \end{array}$$

$$A = \begin{bmatrix} 0 & 0 & 0 & 0 & 0 & 0 & 0 & 0 & 0..37 \\ 0.48 & 0 & 0 & 0 & 0 & 0 & 0 & 0 & 0 \\ 0 & 0.84 & 0 & 0 & 0 & 0 & 0 & 0 & 0 \\ 0 & 0 & 0.84 & 0 & 0 & 0 & 0 & 0 & 0 \\ 0 & 0 & 0 & 0.91 & 0 & 0 & 0 & 0 & 0 \\ 0 & 0 & 0 & 0 & 0.91 & 0 & 0 & 0 & 0 \\ 0 & 0 & 0 & 0 & 0 & 0.95 & 0 & 0 & 0 \\ 0 & 0 & 0 & 0 & 0 & 0 & 0.95 & 0 & 0 \\ 0 & 0 & 0 & 0 & 0 & 0 & 0 & 0.95 & 0.92 \end{bmatrix}$$

Figure 5.13 Projection matrix for the Laysan albatross. The life cycle involves adults that produce eggs and seven one-year stages representing different levels of maturity, all of which are nonreproducing.

occur in the absence of poisoning. That difference (after 50 years) is also equal to about 200,000 additional albatross deaths.

The analysis also indicates that cleaning the lead contamination (i.e., reducing mortality rates due to poisoning from 7% to 0) and increasing adult survival by just 1% (by reducing the rate of drownings) would stabilize the population ($\lambda = 1$). However, if lead-induced mortalities are unabated, then stabilizing the population would require increasing adult survival by twice as much—that is, increasing adult survival by 2%, rather than 1%.

This case represents two important lessons. First and more specifically, decontaminating the soil on this abandoned military base is expensive, but doing so can lead to important improvement for the world's population of Laysan albatross. Moreover, because decontaminating the soil is much easier than reducing the rate of adult drownings, it would be wise to decontaminate the soil.

The second lesson is more general. Recall, the lesson of Case 1: for long-lived species, adult survival tends to be more important for λ than vital rates associated with recruitment. That lesson is important, but it does not mean that recruitment is unimportant or that conservation actions focused on recruitment should be neglected.

Case 3: North Atlantic Right Whales

Some of what makes North Atlantic right whales[13] what they are is also what had made them so vulnerable to being harpooned by whalers.[14] Compared to other whales, they tend to feed near the surface and nearer to coastlines. Furthermore, their bodies are 40% blubber—that's a lot even by whale standards. Consequently, their corpses tend to float. Those traits made right whales easier to harpoon and process than other species of whale. As early as the mid-eighteenth century, the abundance of right whales had plummeted. Right whaling was banned in 1938, though some right whaling continued for several decades afterward.

After the right whaling ban, survival rates and abundance were thought to have slowly increased. But their situation remains dire. And, since the 1980s, the right whales' plight has been worsening. The once common North Atlantic right whale is now reduced to just 300–400 individuals, making it one of the rarest mammals in the world.

Right whales are rare enough to make experts think that they have difficulty finding mates across the expansive waters of the North Atlantic. This results in reduced reproduction, which leads to increased rarity and a repeating cycle of further reductions in reproduction, followed by more extreme rarity—a cycle that ends in extinction. In other words, right whales seemed to have declined to the point of experiencing positive density dependence (recall the marmot example of Chapter 3). Because right whales can live to be centenarians, extinction would inexorably pull right whales to oblivion very slowly and over many decades with nothing for humans to do but reflect on how it all came to be.

However, analyses published in 2001 identified what may be a very brief window of opportunity to save right whales. Here's the story.

Right whales are born with rough patches of skin on their heads, the purpose of which is not fully resolved. What is known is that these callosities are great habitat for whale lice, which may be beneficial to whales. The lice colonies make the callosities appear white when viewed from a distance. Because the shape of each whale's callosity is unique, some humans (and some facial recognition software) have learned to recognize individual right whales from photographs.

Two researchers used a database of 10,000 sightings of right whales from over a 16-year period (1980–1995) to infer vital rates (survival from year to year and frequency of reproduction). From those inferences, researchers built and analyzed a stage-based matrix model, based on the life-cycle diagram depicted in Figure 5.14.

Notice how this life-cycle diagram and projection matrix have a subtle feature that we've not yet seen. The life stages are not strictly tied to age. Instead the stages are tied to reproductive status that may change back and forth over the years. That feature means that the matrix has non-zero values in more places than we've seen before. Previously non-zero values were restricted to the top row (recruitment), diagonal (probability of remaining in the same stage for another year), and sub-diagonal (probability of advancing to the next stage). With this example, we can appreciate that a non-zero element can appear at any place in a matrix. But the principle

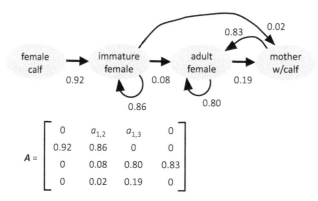

$$A = \begin{bmatrix} 0 & a_{1,2} & a_{1,3} & 0 \\ 0.92 & 0.86 & 0 & 0 \\ 0 & 0.08 & 0.80 & 0.83 \\ 0 & 0.02 & 0.19 & 0 \end{bmatrix}$$

Figure 5.14 Life-cycle diagram and projection matrix for female right whales. Calves are added to the population when an immature female transitions to the mother stage ($a_{1,2}$) or when an adult female transitions to the mother stage ($a_{1,3}$). Arrows representing $a_{1,2}$ and $a_{1,3}$ are not depicted in the life-cycle diagram for the sake of simplicity. This life-cycle diagram is a little trickier than we've previously seen. Notice that an individual can move "backward" from the fourth stage to the third stage with a probability of 0.83.

for interpreting any element is always the same: element, $a_{i,j}$ represents the probability of transitioning from stage j to stage i, except for elements in the top row, which tend to represent recruitment. But as we saw with Case 1, even the interpretation of top-row elements is flexible.

Another distinctive feature of this model is that it attends only to the dynamics of females. This approach is appropriate if the recruitment rate is adjusted to reflect the number of female recruits contributed, on average, by every female of reproductive age. This is an easy adjustment to make if the sex ratio is known.

The researchers' aim was to estimate λ for right whales based on the projection matrix. More precisely, they assessed whether vital rates (and consequently λ) had changed over time and whether estimates of λ indicated population growth or decline. They also used sensitivity analysis to better understand what conservation measures might be required to save right whales from extinction.

Here's what they found. From 1980 to 1995, survival rate had declined, causing λ to decline from about 1.025 (growth) to about 0.978 (decline) (Fig. 5.15). A consequence of reduced survival was that too few females were surviving long enough to reach sexual maturity. More specifically, whereas the life expectancy of females had been more than 50 years in 1980, by 1995, it had plummeted to less than 15 years. Because females are not sexually mature until about age 10 and because they raise 1 calf every 3 to 5 years, the number of calves that a mother could raise in her lifetime fell from 5

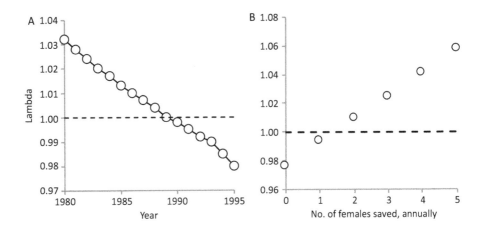

Figure 5.15 Estimated values of lambda for right whales over time (A) and lambda's expected response to saving different numbers of adult females (B). Panel B shows predictions from the projection matrix shown in Figure 5.14. They indicate that saving 2 or more female right whales each year can result in λ > 1, which corresponds to population growth.

to just 1 or 2. Females were finding mates and reproducing frequently. The problem is that their lives were being cut short.

Two key threats to right whales are getting entangled in fishing nets and being struck by ocean-going vessels. Those two threats account for half of all known deaths to right whales since 1970.

Based on that knowledge, researchers took their analysis farther. Because the population has been so small, they assessed how avoiding the deaths of just one or two whales per year would affect λ. Amazingly, saving 1 female per year (from being struck by a ship) increases lambda to 0.995 and saving 2 or more females per year pushes λ past 1.01, which represents population growth (Fig. 5.15). That's all. Spare a few right whales from premature death, and the population can begin to recover.

The analysis binds hope to frustration—hope in knowing that there is a way that might save the whales, frustration because the social forces involved in getting fishing and shipping industries to adjust their behaviors is difficult.

Case 4: Polar Bears

She finds a breathing hole used by seals. She waits. A seal comes for a breath of air. Then the ambush.

For much of the year, polar bears wander across Arctic Sea ice floating over the shallow waters of the continental shelf.[15] They hunt ringed seals and bearded seals, which themselves are foraging for fish.

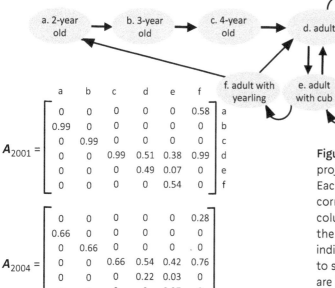

$$A_{2001} = \begin{bmatrix} & a & b & c & d & e & f \\ a & 0 & 0 & 0 & 0 & 0 & 0.58 \\ b & 0.99 & 0 & 0 & 0 & 0 & 0 \\ c & 0 & 0.99 & 0 & 0 & 0 & 0 \\ d & 0 & 0 & 0.99 & 0.51 & 0.38 & 0.99 \\ e & 0 & 0 & 0 & 0.49 & 0.07 & 0 \\ f & 0 & 0 & 0 & 0 & 0.54 & 0 \end{bmatrix}$$

$$A_{2004} = \begin{bmatrix} 0 & 0 & 0 & 0 & 0 & 0.28 \\ 0.66 & 0 & 0 & 0 & 0 & 0 \\ 0 & 0.66 & 0 & 0 & 0 & 0 \\ 0 & 0 & 0.66 & 0.54 & 0.42 & 0.76 \\ 0 & 0 & 0 & 0.22 & 0.03 & 0 \\ 0 & 0 & 0 & 0 & 0.27 & 0 \end{bmatrix}$$

Figure 5.16 Life-cycle diagram and projection matrices for polar bears. Each stage is labeled with letters corresponding to the rows and columns of the matrix. For example, the value 0.54 is the proportion of individuals in stage e that transition to stage f. Values in the top matrix are based on vital rates experienced in 2001 (a year with much ice). The lower matrix is for 2004, which had less ice. Notice how the lower matrix is characterized by lower rates of survival and reproduction.

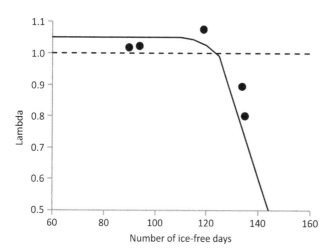

Figure 5.17 Relationship between number of ice-free days and λ. The circles represent each of five years. The dashed line is a reference line (λ = 1), and the solid line is a prediction of the underlying relationship between ice-free days and λ.

Each year when the sea ice retreats from the shoreline, polar bears return to land, where they get by on lesser fare—vegetation, bird eggs, and snow geese. To survive those summer months, polar bears depend on the blubber they accumulated within their bodies from earlier in the year when they feasted on seals.

As the Arctic warms, the duration of each year's ice-free period has lengthened. More precisely, the extent of summer sea ice has been declining by 11% per decade since 1980. This would seem to be bad news for polar bears. Nevertheless, some researchers wondered if there might be a way to assess, more precisely, how that warming would affect the bears' population dynamics.

What these researchers did was estimate the vital rates (survival and reproduction) of polar bears living on the south Beaufort Sea, which lies north of Alaska and northwestern Canada. They estimated vital rates for each of five years between 2001 and 2005. From each year's vital rates and the life-cycle diagram in Figure 5.16, they estimated the corresponding projection matrix and λ.

The researchers found that λ was startlingly lower during years with more ice-free days (Fig. 5.17). Notice that λ is likely <1 (population decline) when the number of ice-free days exceeds ~127. The low value of λ is plausibly due to a combination of factors tied to the increased difficulty of foraging, especially starvation and drowning (as bears are forced to swim greater distances between ice floes).

The researchers went further by predicting polar bear abundances to the year 2100. The predictions took into account how likely each future year is to be a low ice year (>127 ice-free days). For years predicted to have low ice, that year's λ was based on a matrix like that for the year 2004 (Fig. 5.16). For other years, the population's abundance changed according to a matrix like that for the year 2001 (Fig. 5.16). The predictions indicate that polar bears are likely to be extinct or nearly extinct by 2100. This analysis led the US government to declare the polar bear a threatened species under the Endangered Species Act.

ANSWER TO QUESTION ASSOCIATED WITH FIGURE 5.7

$$n_1(t+1) = (0 \times 500) + (3 \times 300) + (6 \times 150) = 1,800$$
$$n_2(t+1) = (0.2 \times 500) + (0 \times 300) + (0 \times 150) = 100$$
$$n_3(t+1) = (0 \times 500) + (0.6 \times 300) + (0 \times 150) = 180$$

CHAPTER 6

Extinction Risk

Human enterprises have increased the rate of species extinction by three orders of magnitude. Consequently, of the ~40,000 species of vertebrates known to inhabit the planet, ~20% are said to be in danger of extinction.[1] Those statistics are the refrain in a dirge for the biodiversity crisis.*

The biodiversity crisis is not only dire; it is also worsening. Species' endangerment has been measurably worsening for several decades. More than half of this worsening is concentrated among just eight nations, including the United States.

While much of this book is a celebration of life's beauty as it percolates through populations, this chapter is steeped in sadness. The situation fosters worry about humans having done too little too late to matter. We can confront somber questions about hope, but not before learning about the population biology of extinction.

Conservation professionals, with their sights set on limiting population extinctions, require an ability to estimate the extinction risk of a population for the purpose of understanding:

· whether a population's extinction risk is high enough to merit special protections.
· what conservation actions would most effectively reduce a population's extinction risk.

---

* Good preparation for this chapter is to quickly reread the opening two pages of Chapter 1.

After discussing the scientific aspects of extinction risk, we'll spend a little time detailing some fraught and eye-opening ethical concerns associated with the science of extinction.

POPULATION VIABILITY ANALYSIS

Metrics of Extinction Risk

Population viability analysis (PVA) is a label that can be attached to any assessment of a population's extinction risk. You already have some experience with PVAs. The case studies involving stage-based models in Chapter 5—about polar bears, right whales, and Laysan albatrosses—qualify as PVAs. Those studies led to estimates of λ, where smaller values of λ were understood to represent greater risk of extinction.

Aside from λ, one of the most important metrics of extinction risk is the probability of extinction over some specified period of time, say, 100 years or some appropriately long period of time. I'll use $P[E_t]$ as shorthand for that metric, where the $P[]$ stands for "probability of" and E_t stands for extinction at time, t. Here are two examples of $P[E_t]$ being used in real-world conservation:

- Scientists working on behalf of the US government have stated that the special protections of the US Endangered Species Act (ESA) should be granted to Hawaiian crows until their extinction risk drops below 5% over a 100-year period and to Steller's eiders until their extinction risk drops below 1% over a 100-year period (US Fish and Wildlife Service 2002, 2009).
- The International Union for Conservation of Nature (IUCN) has developed the Red List Criteria, which are used to classify species according to different levels of endangerment. By this system, a species qualifies as "endangered" if the risk of extinction within 20 years exceeds 20% and "critically endangered" if the risk of extinction within 10 years exceeds 50%.

Those examples beg two questions. First, how does one decide what value of $P[E_t]$ qualifies a population for special protections? Second, how does one estimate $P[E_t]$ for a population? You would be correct if you intuit that answering the first question involves considerable ethical consideration and that the second question is largely scientific in nature. We'll begin with the second question and return to the first question toward the end of this chapter.

Extinction risk can be quantified with a variety of statistics or metrics. One of the most important is $P[E_t]$, the probability of extinction over some specified period of time.

Some Basics

Extinction risk is driven by three properties of a population: average per capita growth rate, temporal variability in growth rate, and abundance. That last property is sometimes distinguished in terms of current abundance and maximum potential abundance (constrained perhaps by habitat). Populations with negative average growth rates are bound for extinction. If two populations with the same negative growth rate differ in abundance, the more abundant population will be expected to persist longer. Populations for which the average growth rate is positive can also be vulnerable to extinction because environmental stochasticity has an average long-term effect of pressing a population toward extinction.[2] Finally, if fluctuations in growth rate increase for a population, so too will its extinction risk. While those basics are straightforward, the details—as we are about to see—are considerably more complex and worth knowing.

Extinction risk decreases with increased average growth rate, increased abundance, and decreased variability in growth rate.

ESTIMATING THE PROBABILITY OF EXTINCTION

A first step in gaining an ability to estimate $P[E_t]$ for real populations is to understand how it can be calculated for a simulated population. Begin by imagining a population with 35 individuals living in a parcel of habitat that can support no more than about 50 individuals. Suppose the population's annual per capita growth rate fluctuates from year to year with an average value of $r = 0.01$/year, and suppose the fluctuations correspond to a variance in r of 0.04. This population can be described by this model:

$$N_{t+1} = N_t + N_t r_t, \tag{6.1}$$

where r_t is randomly drawn, by a computer, from a normal distribution[3] with a specified mean (expected value) and variance, whose values can

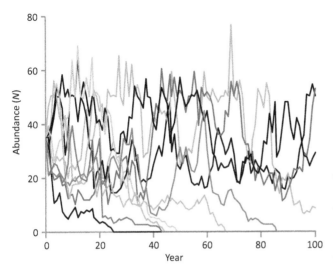

Figure 6.1 Ten stochastic population trajectories created from the model represented by Equation (6.1), with $E[r] = 0.01$, $Var[r] = 0.04$, $N_{max} = 50$ and $N_0 = 35$. Observe that 6 of the 10 populations go extinct within 100 years.

also be represented by the notations $E[r]$ and $Var[r]$, respectively. If the computer picks a random number that would cause N_{t+1} to drop below 2, then we'll say the population went extinct at that time. If the computer picks a random number causing N_t to exceed N_{max}, then an adjustment is made to prevent the population from growing beyond (or much beyond) N_{max}.[4]

The population model is represented not only by Equation 6.1, but also by the rest of the *paragraph* that follows the equation. In this model, N_{max} is a simple way to limit the population's abundance with something akin to the carrying capacity.

We'd run a simulation governed by Equation 6.1 for 100 years and observe whether the population went extinct during that time. Then we'd re-run the simulation many more times. On each occasion, the time series of abundances will be different because the computer will have selected different random values (Fig. 6.1). Do not think of each time series as a precise prediction for what a population is expected to do in the future. Rather, think of each time series as a plausible outcome for what a population might do if it were characterized by those model parameters ($E[r]$, $Var[r]$, N_{max}, and N_0). Because of the model's stochastic nature, there are many, many plausible outcomes. The best understanding of the population's future and its extinction risk comes from considering the collective behavior of these many plausible outcomes.

More precisely we can create, let's say, 1,000 simulated populations and ask what proportion of those populations went extinct. The answer is an estimate of $P[E_{100}]$. The online supplement includes exercises guiding you through the development of an Excel program that does just what I have described here. Working through those exercises is the best way to under-stand what I've just described.

When you perform these simulations, you should get an estimate of $P[E_{100}] \approx 0.34$.[5] That estimated extinction risk depends on the parameters we set for this population (i.e., the parameters given in the legend of Figure 6.1).

Those parameters happen to be loosely representative of the endangered population of ocelots that lived in southern Texas at the beginning of the twenty-first century.[6] Suppose that you wanted to know which of two con-servation actions would lead to a greater reduction in extinction risk for this population. One action would double the amount of suitable habitat, which would cause N_{max} to double. You could rerun the simulations, but this time increase N_{max} from 50 to 100. If you do, you'll find that $P[E_{100}]$ decreases by only a small amount, from ~0.34 to ~0.29.

An alternative action is to reduce the speed limit on roads within the habitat to reduce mortality from car collisions. Doing so might cause $E[r]$ to increase from 0.01/year to 0.035/year. Again, you could rerun the simu-lations, but this time increase $E[r]$ and find that $P[E_{100}]$ decreases greatly, from ~0.34 to ~0.06.

To put a label on it, we've been exploring what is referred to as *count-based PVAs*, so called because the parameters for the model can be estimated from a time series of population counts (i.e., estimates of abundance).

SOME PERSPECTIVE

How can such a simple method for making predictions 100 years into the future be accurate? The question is good, and the answer is complicated. The answer also begins by noting that the accuracy of any forecast derived from a population model—PVAs are no exception—depends on meeting the three assumptions presented in Table 6.1.

Consider assumption 3. Think about how much the natural world has changed over the past 100 years. That in itself gives good reason to think it impossible to reliably predict how model parameters will change over the next 100 years. Nevertheless, concerns about assumption 3 are miti-gated by thinking of $P[E_{100}]$ not so much as a claim about life in 100 years, but rather as an *indicator* of a species' *current* risk. But, if current condi-tions persist into the future, then $P[E_{100}]$ may become a reasonable claim about the future.

I'll discuss assumptions 1 and 2 throughout the chapter, but for now, suppose that the ocelot PVA described above is unlikely to be an accurate

Table 6.1 Assumptions that must be met for an estimate of extinction risk to be reasonably accurate.

Assumption	Comment
1. The population's dynamics are reasonably well described by the population model.*	If, for example, the population exhibits strong negative density dependence at low abundance, then the model represented by equation 6.1 might be inadequate for assuming that dynamics are density-independent except that including N_{max} represents a simple kind of density dependence that may be adequate for some real cases.
2. Model parameters are measured with reasonable accuracy.	Estimates of $E[r]$ and $Var[r]$ tend to be unreliable if they are estimated, for example, from data collected over too short a time span. Adequately reliable estimates of $E[r]$ may require at least 10 to 30 years of data (sometimes more, depending on the required precision). Unless special techniques are used, $Var[r]$ tends to be underestimated for time series shorter than ~50 to 80 years (!) (Ariño and Pimm 1994), and $Var[r]$ can be overestimated if the time series of abundances are measured too imprecisely (Staples, Taper, and Dennis 2004).
3. Model parameters do not appreciably change during the forecast period.	For example, if habitat is lost during the forecast period, then N_{max} will likely decline. Also, the adverse effect of inbreeding (Chapter 7) may cause $E[r]$ to decline over time.

* See Homes et al. (2007) for important nuance about this assumption.

assessment because assumption 1 or 2 is not met. If so, can such a simple method for PVA still be useful?

Without a doubt, yes. The reason is analogous to what we learned in Chapter 2 about the usefulness of density-independent models, even though populations often do not fit the assumptions of that model.

To see more precisely how simple PVAs can be useful, we'll need to run some simulations using an Excel-based PVA. We can build these simulations with the aim of evaluating a simple intuition: extinction risk declines with increasing population abundance. I'll briefly describe these simulations next, and you can find more precise guidance in the online supplement.

STEP 1: The goal is to use Excel to run simulations that result in a graph showing how increases in abundance (x axis) result in changes to $P[E_{100}]$ (on the y axis). For now, we'll set $E[r] = 0$ and $Var[r] = 0.04$. Given our interest in examining the influence of abundance and because we have two parameters pertaining to abundance (N_{max} and N_0), we'll set the initial population size equal to half the value of N_{max}. Then we'll run a set of 1,000 simulations with N_{max} set at 30 individuals and observe $P[E_{100}]$. Then we'll run another set of 1,000 simulations, but this time we'll set N_{max} to 60. We'll repeat that process, running sets of 1,000 simulations for 10 different values of N_{max} (i.e., [30, 60, . . . 300]). The result of these simulations is ten pairs of numbers. The numbers in each pair are a value of N_{max} and the corresponding value of $P[E_{100}]$. These are the values to graph.

STEP 2: Now repeat step 1 to make a new line on the same graph, except this time set $E[r]$ equal to +0.02/yr for this entire batch of simulations. This new line shows how $P[E_{100}]$ declines with increasing N_{max} when the population has a positive average growth rate (as opposed to an average growth rate of 0).

STEP 3: Make one more line on the same graph by repeating step 2. But this time set $E[r]$ equal to −0.02/yr.

> This is a pep talk. I know from experience that making these graphs is time consuming. But repetition is an inescapable key to learning. And if you want to contribute to mitigating the biodiversity crisis as a conservation professional, this is unquestionably worth learning.

Conducting those simulations over an even wider range of parameter values leads to graphs like Figure 6.2, which highlights the influence of N_{max} on extinction risk. To interpret these graphs, take note that the x axes (N_{max}) are presented on a log scale, which makes it easier to see that very large changes in N_{max} are sometimes required for any appreciable change in $P[E_{100}]$.

The y axes ($P[E_{100}]$) are also presented on a log scale, which facilitates observing the effect of changes in N_{max} across salient ranges of extinction risk. For example, reducing $P[E_{100}]$ by half from 80% to 40% would be important, but both values represent high extinction risk. Furthermore, reducing $P[E_{100}]$ by half from 8% to 4% would also be important. Easily seeing both kinds of change is facilitated by the log scale. Continuing that line of thought, there is value in categorizing extinction risk as relatively low (<1%), medium (1% to 10%), and relatively high (>10%). Presenting the y axis on a log scale allows one to assess changes in extinction risk within and between those categories.

Figure 6.2 has three important and obvious patterns: $P[E_{100}]$ deceases with increasing N_{max}, increasing $E[r]$, and decreasing $Var[r]$. Other patterns in Figure 6.2 are less obvious, but no less important and require being

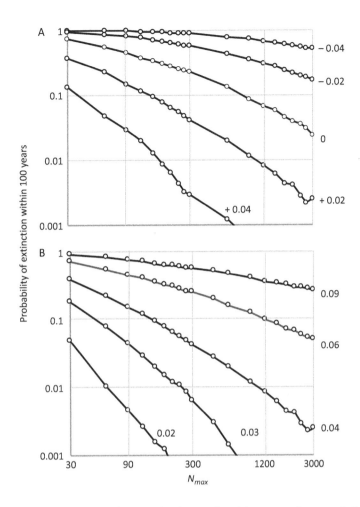

Figure 6.2 The probability of extinction in relationship to N_{max} for populations with different average population growth rate (A) and different variances in growth rate (B). Each circle represents the fate of 10,000 populations simulated according to Equation 6.1. Numeric labels on lines in panel A indicate E[r] for populations represented by each curve. Labels in panel B indicate Var[r] for populations represented by each curve. For all simulations, initial population size was set to N_{max}/2. Var[r] was set to 0.04 for the panel A, and E[r] was set to 0.02/year for panel B.

mindful of the graphs' log scaling. Specifically, large changes in N_{max} do little to reduce extinction risk if E[r] is negative (upper panel) or if Var[r] is high (lower panel). But moderate changes in N_{max} can result in appreciable reduction of extinction risk if E[r] is positive and if Var[r] is low.

The patterns in Figure 6.2 inform real-world conservation. For example, panel A of Figure 6.2 supports the idea that conservation actions leading to more abundance (e.g., habitat restoration or adding individuals from a captive population) are unlikely to matter much if, for example, overexploitation causes the population to have a negative per capita growth rate. But increasing habitat (which would increase N_{max}) is far more likely to make a big difference if the population's vital rates are healthy (i.e., $E[r] > 0$). The idea is intuitive, but perhaps only after having studied Figure 6.2. Similar insight is to be found with panel B of Figure 6.2. Increasing abundance leads to large declines in extinction risk when Var[r] is low, but only small declines in extinction risk when Var[r] is high.

> Given the now-apparent importance of these subtler patterns, I recommend rereading the previous paragraph and confirming the accuracy of those statements by carefully comparing them to Figure 6.2.

It is often not feasible to reduce Var[r] via conservation action. Furthermore, single, isolated populations with high Var[r] are prone to extinction unless they are very large. The full richness of that previous sentence is given as much by the phrase "single, isolated" as by the phrase "very large." To see how, consider that many insect species are habitat specialists, often because they lay their eggs on just one or a few species of plant. That specialization is often associated with insect populations occupying small, isolated habitat patches.

As a result, the N_{max} for many insect populations is small. Insect populations are also often characterized by large fluctuations in abundance (i.e., high Var[r]). That combination of low N_{max} and high Var[r] means that insect populations living on small patches of habitat are extremely vulnerable to extinction. While these patches are isolated, they are often close enough to each other that an empty patch can be recolonized by individuals dispersing from a nearby occupied patch. This network of patches and subpopulations is called a *metapopulation*. While a subpopulation may not last long, the metapopulation may experience low extinction risk.

Insects are not the only species that frequently persist because of metapopulation dynamics. It turns out that many vertebrate species have extinction dynamics much like what I just described for insects. More precisely, Fagan et al. (2001) used count-based PVAs to conclude that extinction dynamics can be placed into one of three broad categories, based on the values of Var[r] and r_{max} that describe a population.[7] These categories are as follows:

· A: populations that have low extinction risk even when abundance is low.

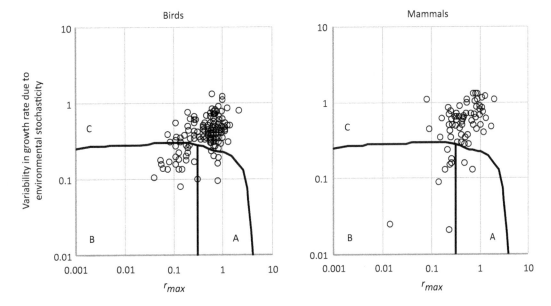

Figure 6.3 Populations of birds and mammals analyzed by Fagan et al. (2001), each of which is characterized by r_{max} (x axis) and variability in r (y axis). Count-based PVAs were analyzed to determined that different combinations of r_{max} and variability in r result in different categories of extinction dynamics, represented by the labels A, B, and C and described in the main text. This figure is adapted from Fagan et al. (2001), who also found similar patterns for species of fish and Lepidoptera.

- B: populations whose extinction risk declines significantly with significant increases in abundance.
- C: populations with high extinction risk, even if abundance is high.

For species belonging to that last category, persistence depends on meta-population dynamics or existing over very large geographic ranges such that the adverse impact of environmental stochasticity (i.e., bad years) affects different parts of the range in different years. The technical term for this is *spatial asynchrony*.

That tripartite categorization is depicted in Figure 6.3. Each open circle on those graphs represents a different time series—several hundred in all, representing various species of birds and mammals. The graphs represent each population in terms of its r_{max} and Var[r].

The labeled regions of each graph (A, B, C) correspond to the aforementioned categories of extinction dynamics. In particular, region A represents cases where r_{max} is high and Var[r] is low. Region B represents cases where r_{max} and Var[r] are both low. And region C is primarily characterized by cases where Var[r] is high. Those combinations of high and low r_{max} and Var[r] are what give populations the extinction dynamics described two paragraphs before. You'll gain additional insight by studying Figure 6.2 until you see that the patterns there foreshadow the connection between extinction dynamics and population parameters being explained here. Finally, note that very few populations belong to category A and most populations belong to category C (Fig. 6.3).

Count-based PVAs can be readily modified to accommodate more realistic model structures. For example, it would be easy to replace Equation 6.1 with an equation representing linear negative density dependence (Equation 3.3) or some other form of density dependence, such as θ logistic (Equation 3.4) or positive density dependence. Taking adequate account of density dependence is important, because negative density dependence operating at low abundance tends to reduce extinction risk and positive density dependence at low abundance tends to exacerbate extinction risk.[8]

Some populations live in environments where environmental stochasticity is autocorrelated, meaning that good years tend to occur consecutively, as do bad years. Again, the count-based PVA that you built in Excel can be readily adjusted to represent such dynamics. Taking account of autocorrelation when it occurs is important because it tends to exacerbate extinction risk (Ripa and Lundberg 1996).

Determining r_t by drawing numbers at random from a normal distribution (as done for the model associated with Equation 6.1) is a useful way to account for "run-of-the-mill" good years and bad years. But that approach is not good at representing rare catastrophic events, such as so-called 1-in-1,000-year weather events or episodic outbreaks of disease.

These more exceptional events were examined by Reed et al. (2003a), who studied the time series of abundances for populations representing 88 different species. They found that catastrophic declines of 50% or more occur with a probability of 14% per generation. That corresponds to about once every 20 years for creatures with a generation time of 3 years and once every 50 years for creatures with a generation time of 8 years. The severity of die-offs also varied greatly, ranging from 50% to ~95% declines in abundance. While 90% dies-offs were quite rare, die-offs of up to 75% were relatively common. Although catastrophes are known to be important, modeling their effect on real populations is often more guesswork than not, given that wide range of severity.

The modifications to the model structure discussed thus far can all be applied to count-based PVAs. While count-based PVAs can be built in Excel, real-world applications tend to be built using computer programming platforms such as MATLAB, R, Python, Java, C, or C++.

Aside from count-based PVAs, another basic structure for PVAs is the stage- or age-structured population model. These models are like those discussed in Chapter 5, except they tend to account for stochastic vital rates. That is, the individual elements of the transition matrix are drawn randomly from probability distributions that represent fluctuations in vital rates (age-specific fecundities and survival rates) due to environment stochasticity. There is also typically a need to account for within-year correlations among vital rates (i.e., a bad year for one vital rate is often a bad year for other vital rates; Fieberg and Ellner 2001). The data required to estimate interannual variability in vital rates and correlations among vital rates are hard-earned and often not available.

Another common structure for PVAs is an *individual-based model*, where a computer keeps track of the fate of each individual of the population. This is the most flexible type of PVA—able to accommodate a wide range of ecological and genetic detail. Some individual-based PVAs are built specifically for a particular case. But there are also a number of readily available software packages that estimate extinction risk using individual-based PVAs (Lindenmayer et al. 1995). Examples include VORTEX (Lacy, Miller, and Traylor-Holzer 2021) and HexSim (Schumaker and Brookes 2018). Such software works by a user inputting all the parameters necessary for making an estimate of extinction risk, such as age-specific vital rates. The parameters are entered into the computer via the software's intuitive interface. The details of these different programs vary importantly, but a useful sense of how they work follows.

The simulated populations may begin with 100 individuals, each characterized by some set of properties (e.g., sex, age, genotype or genetic relatedness to other individuals in the population, socio-reproductive status, such as being a dispersing or reproductive individual). Each year, each individual survives or dies with a probability that depends on its age, sex, and so on. Vital rates can be modeled to improve or deteriorate over time. For each year of the simulation, the program determines which individuals reproduce, giving rise to new individuals to track in the upcoming year.

Some software for individual-based PVAs allows for explicit modeling of spatial dynamics, where the subpopulation or habitat patch to which each individual belongs is tracked, and vital rates depend on the quality of the habitat patch.

These programs calculate abundance each year and record when the population goes extinct. Thousands of populations are simulated. The fate of those populations is the basis for reporting various PVA statistics such as the $P[E_t]$ or the mean (or median) time it takes for a population to go extinct. While the details of an individual-based PVA are complex, the basic idea is similar to the Excel program that you built.

To develop a better understanding for how individual-based PVAs work, the online supplement includes an exercise guiding you through the methods and results of some published research that assesses extinction risk via individual-based models.

UNIVERSAL MINIMUM VIABLE POPULATION

To this point, I have presented, or at least pointed to, a rich repository of concepts and tools for building useful PVAs for a wide range of species. Those tools and concepts contribute to the ability to build reliable PVAs that at least aim to meet assumptions 1 and 2 of Table 6.1. Nevertheless, the data and expertise required to build such PVAs are often not available.

For example, scientists working with the IUCN have assessed extinction risk for as many of the planet's species as is possible. By their records, about 15% of the world's ~5,000 species of mammals are known by so little data as to make it impossible for scientists to determine whether they are vulnerable to extinction. Among amphibians the proportion of data-deficient species is about 24%.

More evidence for the limits imposed by inadequate data and expertise comes from Chaudhary and Oli (2020). They developed a list of 32 properties that an excellent PVA should have. These 32 properties are essentially a more detailed accounting of the assumptions in Table 6.1. They compared this list of properties to 160 PVAs published in the scientific literature between 1990 and 2017. Of those PVAs, they considered only 18% to possess enough of those 32 properties to be considered adequate.

Because the data and expertise required to perform a proper PVA are so commonly unavailable, there has been keen interest to discover a smallest population size that tends to result in low extinction risk for most species. If such a number could be discovered from the many well-studied populations, then maybe that number could be usefully applied to cases where data or expertise are unavailable. This holy grail of a number has been referred to as a universal *minimum viable population size* (MVP).

Researchers have searched for a universal MVP by reviewing the results of many hundreds of PVAs. One study defined an MVP as the smallest abundance that would result in a probability of extinction over 100 years that is no greater than 10% (Brook et al. 2006). The researchers estimated this smallest abundance from count-based PVAs parameterized from about 1,200 time series. The median MVP was about 1,400 individuals. However, MVPs commonly ranged from as low as 400 to as high as 150,000 individuals. Another study focused on PVAs built from VORTEX that also accounted for the effect of inbreeding depression (Reed et al. 2003b).[9] That study also observed that MVPs varied greatly among populations and that the median MVP was ~6,000 individuals.

Why do MVP sizes vary so much from one population to the next? Important insight can be found in the two graphs of Figure 6.2. For each of those graphs, draw a horizonal line at $P[E_{100}] = 0.10$. Observe how that horizontal line crosses the five curves in each graph at different values of N_{max} (or, would cross the curves if the x axes of Figure 6.2 were extended farther to the right). Those values of N_{max} are MVP sizes for populations of given $E[r]$ and $Var[r]$. In other words, MVP size varies so greatly because it is very sensitive to changes in $E[r]$ and $Var[r]$. MVP size also varies among species depending on the nature of density dependence, sensitivity to inbreeding, spatial and temporal autocorrelation in vital rates, and more.

Many have interpreted those results as meaning there is no universal MVP size. I agree. However, the results are still very useful: in the absence of information to the contrary, a population of 50,000 individuals is probably not at great risk of extinction, a population of 5,000 individuals might be in danger and it might not be, and a population of a few hundred individuals is probably in danger. I realize that's wildly imprecise, but it's not nothing.

TAKING STOCK

Let's review what seem to be two key lessons to this point. First, a population's extinction risk tends to decrease as its abundance increases, but the details of that relationship vary greatly among populations (even among populations of the same species). Much (certainly not all) of this variation can be accounted for by considering three kinds of populations:

- Populations with low variability and high growth rate can persist for long periods at low abundance. Think, for example, of populations whose MVP is on the order of a few hundred. These are also the populations in region A of the graphs in Figure 6.3. Few populations fit this description, and it would be unwise to presume that any particular population of conservation concern fits this description unless compelling evidence indicated as much.
- Some populations experience significant reductions in extinction risk with significant increases in population size. They may be viable once their abundance reaches a few thousand or so. Some populations fit this description. These are populations in region B of the graphs in Figure 6.3.
- Most populations experience significant reductions in extinction risk only by some combination of being abundant, occupying large

geographic ranges, being part of a metapopulation, and the amelioration of threats that limit population growth rate (E[r]). These are populations in region C of the graphs in Figure 6.3.

A second key lesson is that it is often not feasible to build a PVA that reliably indicates the category (A, B or C) to which any particular population belongs, let alone produce a satisfyingly accurate estimate of extinction risk. For such populations it would be extremely valuable if there were a universal MVP. There isn't.

THREATS, ACTIONS, AND OUTCOMES

The impression left by those lessons is bleak and important, but also incomplete. A more complete understanding requires additional considerations. First, suppose that a PVA is a tool not merely for estimating extinction risk, but also for developing action plans that reduce extinction risk. If so, the purpose of a PVA can be to link threats, actions, and outcomes—that is, to link *threats* faced by a species to *actions* expected to mitigate the threats and *outcomes* expected to result from the actions, where the outcomes are considered in terms of reduced extinction risk (or improvements in the determinants of extinction risk, such as abundance and population growth rate). This conceptually simple process can be applied both to cases in which much is known about the species and to those in which uncertainty abounds.

Desert tortoises of the Mojave Desert in the southwest United States are a species wrapped in considerable uncertainty (Darst et al. 2013). They are long-lived creatures whose basic life history is known well enough to build a useful stage-based population model. But desert tortoise conservation has long been stymied by not knowing well enough how to connect desired *outcomes* (i.e., reduced extinction risk) to conservation *actions* that mitigate *threats* to the species. One challenge is that desert tortoises experience a diversity of threats, and the relative impacts of threats has been difficult to estimate.

Darst et al. (2013) identified 44 different threats to desert tortoises, ranging from off-road vehicles (that sometimes crush tortoises) to ravens (that prey on juvenile tortoises) and landfills (that favor increased abundance of ravens). The impact of those threats on tortoise population dynamics was analyzed with a stage-based population model, where the impact of the threats on vital rates (survival, reproduction, and population growth rate) was estimated by asking experts their opinion of how each threat was related to tortoise vital rates. From the list of 44 threats, 3 were identified as likely being responsible for about half of the total risk to desert tortoise population dynamics. Those top-ranked threats were urbanization, military operations, and civilian access to sensitive habitat.

The point of this example is to show that the principles of PVA can be used to guide conservation even in cases in which the specific data necessary for building a formal model are unavailable.

This threats-actions-outcomes approach has also been applied to northern spotted owls in the Pacific northwestern region of the United States. In that case, the challenge was to understand what actions would most mitigate the combined threat of habitat loss and barred owls (which can outcompete spotted owls). You can learn more about this case by reading Dunk et al. (2019).

ADAPTIVE MANAGEMENT FOR EXTINCTION RISK

Let's approach that sequence of relationships (threats-actions-outcomes), from another perspective. Suppose there is a societal willingness to take necessary actions to conserve a population, but too little information to know—with enough confidence or precision—what actions should be taken because the threats are poorly understood. By "societal willingness," I am referring to the funding, public support, and political paths required to implement meaningful conservation actions.

If knowledge for precisely how to conserve a species is limited, but adequate willingness is present, then PVAs and the conservation planning they support may be incorporated into a kind of *adaptive management*, whereby a PVA suggests a set of conservation actions, those actions are implemented, their effects are observed and evaluated in a revised PVA, and the conservation actions are modified accordingly. At this point the process comes full circle and begins again. If that process begins from a position of very limited knowledge, it can adopt provisional assumptions, such as these:

1. Assume that reducing extinction risk (the ultimate desired *outcome*) requires *actions* that increase abundance (N_{max} and perhaps N_o) *and* population growth rate (E[r]), unless and until evidence suggests that increases in only one of those basic parameters is required. This assumption is supported by Figure 6.2.
2. Assume that there are multiple *threats* and a significant risk of misjudging the relative importance of the threats. An example includes the possible over-emphasizing of raven predation on juvenile desert tortoises in the Mojave Desert and underemphasizing habitat destruction (see also examples in Chapter 5).

3. For longer-lived species, assume population growth rate is especially sensitive to adult survival. For shorter-lived species, assume that population growth rate is similarly sensitive to survival and recruitment of juveniles. This assumption is supported by Figure 5.12. If the species' longevity is not known, it can be estimated from knowledge of a better-studied and closely related species[10] or on the basis of body size.[11]

4. If the population's abundance is in the hundreds, achieve an increase in abundance to the thousands as quickly as possible. If the population's abundance is on the order of a few thousand individuals, work diligently to increase that abundance to a few tens of thousands. Before any such target is reached, you'll likely have the time required to gain enough information to set goals for abundance on the basis of a more detailed and case-specific PVA.

5. Establish more than one population to distribute—rather than concentrate—the risk of a catastrophe. An illustrative example is the establishment of two whooping crane populations (in Texas and Louisiana). While a hurricane could decimate one population, the other population would be a safe distance away. Another example is the red-cockaded woodpecker, which lives in scores of small, separate populations that are concentrated mostly in eastern North Carolina and vulnerable to being wiped out by a single hurricane. An appreciable reduction in extinction risk requires establishing more populations outside the path of a single hurricane.[12]

Those assumptions are merely a provisional starting point, subject to revision if available information (or new information) indicates the need for revision. For example, those aforementioned initial assumptions would be revised for Mojave Desert tortoises on the basis of analyses like that described in the section that immediately precedes this one.

For poorly studied populations, revision to those assumptions would result from learning what has been threatening the species all along. For well-studied species, revision would result if one were to learn that threats are changing over time. An important example of that second circumstance occurred with the northern spotted owl. Early in the conservation planning process, during the 1980s and 1990s, the single major threat was habitat loss due to overlogging old- growth forests. Then a second major threat emerged during the first two decades of the twenty-first century as the abundance of barred owls increased. The problem caused by barred owls is that they tend to reduce the vital rates of northern spotted owls through interspecific competition—primarily, what seems to be interference competition.[13]

WILLINGNESS AND ABILITY

Adaptive management for extinction risk may be the most reliable way to reduce extinction risk when there is limited knowledge but ample societal willingness. Unfortunately, societal willingness has too often been lacking. One of many possible indications of inadequate societal willingness are decision-makers thinking that some set of actions required to appreciably reduce a population's extinction risk are too ambitious.

A stronger and broader indication that conservation action is limited more by societal willingness than knowledge arises from the IUCN's assessment of extinction risk for mammals. Many such assessments have been made at two points in time (1996 and 2008) for the same species (Hoffman et al. 2011). Among those species, many did not change extinction risk categories (vulnerable, endangered, critically endangered; Table 6.2), some changed to a higher risk category, and others changed to a lower risk category. Twenty-four of the assessed species were reclassified into a less endangered class (e.g., from endangered to vulnerable, Table 6.2), and 171 species were reclassified into a more endangered class (e.g., from endangered to critically endangered). To put a point on those numbers, deteriorations were seven times more common than improvements. Of the 171 deteriorations, 85% were cases in which necessary conservation actions were not taken even though the knowledge and technical ability to do so had been available. Only 15% appear to have been cases in which conservation action was stymied by lack of knowledge or technical ability. An example of being technically unable to mitigate extinction is that of North

Table 6.2 IUCN Red List Criteria. By these criteria, species are placed into one of three broad categories of extinction risk. This table is a partial summary of those criteria. The complete set is readily available on the internet and includes additional criteria and more precise descriptions of the criteria listed here.

Criterion	Vulnerable	Endangered	Critically endangered
Abundance	<1,000	<250	<50
Reduction in abundance	≥50%	≥70%	≥90%
Geographic range	<20,000 km^2	<5,000 km^2	<100 km^2
Extinction risk	≥10% in 100 years	≥20% in 20 years or 5 generations	≥50% in 10 years or 3 generations

American bats affected by white-nose syndrome: we know disease is the threat, but do not know how to treat or prevent it.

Changes from one IUCN risk category to the next are quantum leaps toward or away from extinction. Most of the species considered in the analysis that I just described did not change categories between 1996 and 2008. Thus, there is value in considering data that might speak to changes at a finer scale. For example, a team of 89 researchers compiled estimates of population decline for 32,000 populations representing 5,320 species of vertebrates and found that, on average, those populations declined by 69% over the past 50 years (1970 to 2022) (World Wide Fund for Nature 2022). A similar study—focused on mammals and published more than a decade earlier—found similar results (Collen et al. 2009; Boitani and Rondinini 2010). Many of these populations that have been studied well enough to estimate rates of decline over a 50-year period would also be studied well enough to apply adaptive management for extinction risk. Observing such declines raises concern that these declines are at least partially (if not largely) fueled by widespread lack of societal willingness.

CONSERVATION TRIAGE

Widespread lack of societal willingness is manifest, in part, through widespread shortfalls in funding for conservation. That circumstance leads some conservation professionals to advocate for allocating those inadequate funds in a manner that proactively decides which aspects of biodiversity *should* be conserved and which *should* be forsaken in the belief that such allocations of funds will save more biodiversity than other allocations. This idea is known as *conservation triage*. I italicized "should" to remind us that triage is an ethical decision and that the ideas of Chapter 4 apply to this topic.

Some advocates of conservation triage specify the need to calculate the efficiency (E) of alternative conservation actions and allocate funds based on that efficiency, which is represented by this formula (Bottrill et al. 2008):

$$E = (V \times B \times S)/C, \qquad (6.2)$$

where V is the overall value of the particular object of conservation concern, B is the benefit of the action to that particular object of conservation concern (say, an endangered species), S is the probability of successfully realizing the hoped-for outcome, and C is the cost of the action. The idea, in part, is to implement the most efficient actions that can be afforded.

Other advocates argue, instead, that conservation triage should be guided by withholding funds from species with especially high extinction risk (because they are doomed) and relative low extinction risk (because they are in less need) and focusing funds on species with intermediate levels of extinction risk.

Some argue that conservation triage is merely strategic decision-making because some species will be forsaken—no matter what—so it is better to forsake proactively according to some preplanned prioritization. Others argue that conservation triage is deeply unstrategic in the belief that it signals to decision-makers that valuers of biodiversity tacitly accept (or do not object to) sociopolitical circumstances that yield inadequate funds. The effect of that signaling—according to those holding this objection to conservation triage—is to further limit the prospect for securing funds.

Conservation triage deserves our consideration for a couple of reasons. First, PVAs play a critical role in conservation triage. They do so by extending our PVA framework (threats-actions-outcomes) to also taking account of the cost of actions. Second, the appropriateness of conservation triage is unresolved, and evaluating its appropriateness is an important ethical issue arising from the assessment of extinction risk.

To enrich our understanding of conservation triage, let's begin with some examples.

Kit Foxes and Sumatran Tigers

When an endangered species remains as a set of more-or-less isolated populations due to habitat fragmentation, triage planning entails asking how inadequate funds should be allocated among the populations of a species to get the greatest reduction in overall risk of extinction for the species. Should available funds be divided evenly among the populations (and risk not making a substantive improvement to any population)? Or, should all funds be focused on one population (and risk neglecting the other populations)? Or, is there an intermediate strategy that works better?

Advocates for conservation triage developed an equation for such a case showing that there is no single answer to the question (McDonald-Madden, Baxter, and Possingham 2008). They also show that the best answer is not necessarily intuitive and depends on the precise relationship between funding allocation and risk of extinction for an individual population. That relationship is not known for the vast majority of cases. But these researchers found two cases in which they could apply their formula—San Joaquin kit foxes (a subspecies of kit fox) and Sumatran tigers (a subspecies of tiger).

Most of these kit foxes live in one of eight populations in the San Joaquin Valley of central California. An important strategy for reducing their extinction risk is to protect and expand habitat by purchasing private land. Most Sumatran tigers live in one of four populations on the island of

Sumatra in Indonesia. An important conservation action for these tigers involves the allocation of funds to reduce poaching in each population.

The researchers estimated the optimal allocation of funds to each population for those cases. The optimal allocation is complicated and not readily described in the space available here. But the more important result—emphasized by the researchers—is that for the assumed budgets ($10 million for foxes, $52,000 for tigers), the mean time to extinction for each subspecies was less than 32 years. The researchers also estimated how much extinction risk is reduced with increased budgets.

I have been uncertain about an important aspect of this research. I would have expected an advocate for conservation triage to conclude that all funding be removed from Sumatran tigers and San Joaquin kit foxes because the overall prospects for their survival are so bleak. However, that consideration was not raised, let alone evaluated. As such, this research has impressed me as being less about conservation triage and more about the importance of cost-effective conservation action and the determination that funds are presently inadequate (with some sense for how much funding would be required).

The US Endangered Species Act

In 2016, the US Fish and Wildlife Service was legally obligated to work toward the recovery of the 1,124 species protected by the US Endangered Species Act (ESA). To fully fund the budgets estimated as being required for making good progress toward recovery would have cost $1.21 billion per year. But only ~25% of those funds had been made available by the US Congress.[14] Given the overall shortfall in funding, most individual recovery programs are underfunded. Meanwhile, recovery programs for some species have spent more than was budgeted. Those circumstances inspired the analysis presented in Gerber (2016) and summarized here.

For each endangered species, it was determined whether more or less funds had been spent in relationship to what had been budgeted. For the same species, it was also estimated whether the species tended to increase or decrease in abundance during the 22-year assessment period, 1989–2011. One key result was that "overfunded" species were more likely to experience increases than underfunded species. A second key result was recognition that a number of programs had failed to result in improvements to the species, but were especially over budget.

This analysis continued with what its author describes as a "thought exercise." For this exercise, conservation triage was defined narrowly as the "reallocation of surplus spending from the top 50 costly but heretofore futile recovery efforts to efforts that are grossly underfunded." Those 50 cases were labeled "costly failures." The "excess" funds from those "costly failures" was enough to erase the budget shortfalls for 182 species that had been neglected. The author of this analysis concluded that "in light of the high-risk level and potentially unexpected consequences associated with triage, . . . the results should not be used to definitively identify species for triage until more data allow for validation of the quantitative relationship between funding and [conservation] success" (Gerber 2016, p. 3565).

I agree. For me, the analysis is valuable for exposing questions about why some over-budget projects were seemingly ineffective. Were they improperly budgeted in the first place? Had too little time passed to expect an appreciably improved status? Were they properly budgeted, but in need of better management that would not have required more funding?

Those questions and the analysis that inspired them all speak to the need for identifying and pursuing cost-effective conservation. But it is not quite clear how they inexorably lead to proactive decisions to forsake some species.

Woodland Caribou

Woodland caribou are severely threatened across Canada by habitat loss and wolf predation. Both are the result of oil and gas extraction.[15] If recent trends continue in Alberta, Canada, then most caribou herds there will be reduced to fewer than 10 individuals by 2040. Advocates for conservation triage estimated the cost of conserving the 12 herds that represent most of the woodland caribou in Alberta (Schneider, Hauer, and Boutin 2010) by taking account of the costs of killing wolves (to reduce predation on caribou), accelerating the reforestation of habitat ruined by oil and gas extraction, and paying the gas and oil industry for the profits that would be foregone if caribou habitat was protected. The estimated cost for protecting each herd varied due to its size, existing damage, and projected value of the land to industry. The estimated cost of effectively conserving a herd ranged from $19 million to $30 billion (Canadian dollars) in 2010.[16]

These advocates for triage refrained from saying which herds should be forsaken or even which should be first to be forsaken. Their doing so is noteworthy because deciding what to proactively forsake is the essence of conservation triage. Without making that call, the assessment most impresses me as a dire warning, as opposed to justifying conservation triage, let alone an actual instance of triage planning.

If Equation 6.2 were the basis for triage, an advocate for triage would have to assign numbers to the parameters of that equation. Doing so would require robust answers to questions such as the following: Should B (from Equation 6.2) represent the *limited* benefit to the species of protecting a particular herd? Or should *B* represent the much *broader* benefits of curtailing oil and gas extraction? Limiting *B* to the benefits for caribou risks myopic application of triage. In any case, decisions about how to handle *B* are not obvious and would matter greatly.

Equation 6.2 also requires replacing *V*, the value of woodland caribou, with a number. Should *V* be the value of caribou to environmentalists or to the oil and gas industry? Perhaps *V* should be the overall value to human well-being? I'm not sure how one makes those decisions, or how one replaces *V* with a number. Nevertheless, those decisions also matter greatly to the outcome.

For this case and many others, it seems unwise to presume that the parameters that determine the outcome of triage can be reliably replaced with numbers.

A Grassland Ecosystem

Only 30% of Canada's semi-arid, mixed-grass prairies remain intact. An important parcel of this ecosystem type is the Milk River Watershed (~1.5 million ha) in southern Saskatchewan. The watershed is inhabited by nine endangered species, including burrowing owls, greater sage grouse, and swift foxes. Advocates for conservation triage aimed to determine the most cost-effective suite of conservation actions from among 14 sets of potential actions (Martin et al. 2018). The sets of action were wide ranging and included, for example, habitat restoration, limiting human activities in sensitive habitat, and encouraging landowners to modify the use of their land. The researchers convened a group of experts for a three-day workshop to elicit their beliefs on the financial costs (*C*), conservation benefits (*B*), and feasibility (*F*) of various sets of conservation action. Feasibility was much like *S* in Equation 6.2 and was defined as experts' collective belief about the probability an action would be technically possible and socially acceptable. Benefits were defined as experts' collective belief about the improved probability of realizing recovery goals for each of the species, given successful implementation of the actions. The experts also decided on interim (20-year), site-specific (limited to the watershed) recovery goals. On the basis of those goals and beliefs, the researchers estimated which set of actions would result in the greatest probability of meeting the most recovery goals.

They concluded that taking account of all three factors (cost, benefit, and feasibility) typically led to better conservation outcomes than taking account of only the cost or only the benefit of actions. They also found that increasing budgets led to better chances of realizing more recovery goals. For the largest budget considered (~$180 million, Canadian dollars, over 20 years), only three species had a 70% or greater chance of realizing the recovery goals.

Two features of this analysis most catch my attention. The first feature is the quantification of C, B and F by elicitation of expert opinion, which is sometimes the best-available method for quantifying parameters in a model. While the authors of this analysis used a high-quality method for eliciting expert opinion, the quality of methods used for eliciting expert opinion vary greatly from case to case. Poorly elicited opinions can be terrible, the expertise and effort required for high-quality elicitations tends to be under-appreciated, and even the best elicitations can be inadequate.[17] As such, important unanswered questions include: How frequent are cases for which currently available knowledge of C, B and F are too limited to reliably support decisions to proactively forsake some particular recovery goal? How does one identify those cases?

Second, the conceptualization of B in this case raises questions about the scale and scope of triage. Recall that B was focused on meeting interim recovery goals on the (small) portion of species' ranges within the Milk River Watershed. It could be a mistake to assume that the risk of failing to meet interim, site-specific goals represents reason to proactively forsake actions supporting such goals. Adequate triage planning would seem to require assessing what is happening on the remaining portions of species' range. Those considerations make me think this analysis is another case in which maximizing the efficiency of expenditures is conflated with conservation triage.

Unanswered Questions

My sense is that the preceding account is representative of what advocates have to say about how and why conservation triage should be performed. As such, it seems to me that they have provided genuinely valuable insight for understanding cost-effective conservation and for quantifying how wildly underfunded and undervalued conservation is.

But the essence of conservation triage seems well beyond the pursuit of cost-effective conservation. It seems to be essentially about deciding what species or outcomes *should* be proactively forsaken. Those decisions seem to require at least three determinations that have not yet been adequately provided:

1. Conservation triage seems to depend on the algorithm used for deciding which species (or conservation outcomes) to forsake, but there is no agreement on which algorithm to use. One possibility is Equation 6.2. Another is the algorithm used in the grassland

ecosystem example, which does not take account of value (*V*), the effect of which is to implicitly assume that *V* is equal for all conservation goals. Another possibility—advocated by some, though not fully explicated—is to withhold funding from species or populations with the highest risk of extinction (because they are doomed to extinction) and the lowest risk (because they are in less need of conservation) and focus funds on populations with intermediate risk of extinction, because those populations would benefit most (e.g., Arponen 2012). A well-qualified group of researchers have advocated using the "SAFE index" as a basis for deciding what to forsake (it essentially forsakes species whose abundance is well below 5,000 individuals) (Clements et al. 2011).[18] Other well-qualified researchers have vigorously criticized that approach (McCarthy et al. 2011). The US Fish and Wildlife Service uses another algorithm for prioritizing which species to protect under the Endangered Species Act (Alexander 2010). The lack of broad agreement about what triage algorithm to use is concerning.

2. Once an algorithm is selected, parameters of the algorithm need to be assigned a numerical value. Obstacles to doing so include some assignments depending greatly on normative judgments and some assignments requiring empirical data that is typically not available.

3. Conservation triage seems to require a judgment about the most appropriate scale or scope of application. Should it be applied among populations within a species (as with kit foxes and Sumatran tigers) or among species (as with the grassland ecosystem and ESA examples)? Should it be applied to alternative objects of conservation (this species or that) or to alternative actions (such as whether or not to curtail a broad threat that affects many species)? All scales of conservation are important, but applying conservation triage at one scale seems to risk being at cross-purpose with other scales of conservation. For example, conservation triage applied among the populations that compose species A would implicitly encourage spending funds on species A, but conservation triage applied to a broader set of species might indicate that species A should be forsaken.

Is Conservation Triage Strategic?

If conservation triage were a private conversation among conservation professionals, perhaps concerns would be limited to those mentioned above. But that is not the case. Conservation triage is also public discourse between conservation professionals and the rest of society.

Consequently, I worry about how conservation triage affects, for example, the attitudes of high-level decision-makers who have the most impact on conservation, but know the least about conservation (i.e., business leaders and elected officials). Their knowledge about conservation—including the value (V) of various conservation goals and the harms associated with failing to conserve—comes from conservation professionals.

I also worry that conservation professionals arguing to proactively forsake some conservation goals are unwittingly signaling to uninformed decision-makers that those conservation goals are not so valuable. Could the more strategic action be focusing all discourse on explaining the value of conservation, especially in relationship to competing values such as freedoms to pursue livelihoods that threaten biodiversity? You'll recall from Chapter 4 that the conservation profession has plenty of room to improve skills for evaluating and explaining conservation values (V) to other segments of society—especially given that too many decision-makers see conservation as no more than one of many ideologies or special interests to navigate in a political world.

If triage has some appropriate, albeit limited, role to play but also works against efforts to explain the value of conservation, then how should conservation professionals most strategically allocate efforts between advocating for triage and advocating for conservation's value? I don't think anyone knows. Why? Because no one really knows the chances of realizing substantive change in societal willingness to better conserve biodiversity or how much time we need (or have) to make such changes.

The Risk of Faux Triage

I also worry that discourse about conservation triage makes it easier for decision-makers to curtail conservation for reasons that have less to do with funding and more to do with political expediency. Red wolves in the southeastern United States illustrate the concern. The legal responsibility for recovering red wolves rests with the US Fish and Wildlife Service (FWS). In 2016, the FWS appeared to retreat from conserving red wolves in the wild, the rationale being "maximizing efficient use of Service resources" (US Fish and Wildlife Service 2016). But the FWS never provided any assessment of efficiency. Moreover, it is at least plausible that its retreat was the result of political opposition to wolves by local landowners and state governments (Fears 2016). If so, the retreat was driven not so much by concern for conservation efficiency, but an unwillingness to withstand political pressure. Then the FWS covered that explanation with the language of conservation triage. Meanwhile disagreement remains enflamed among segments of society over the value (V) of a wild population of red wolves.

Inappropriate Analogy

Because conservation professionals communicate and act in public, I have raised the concern that conservation triage miscommunicates the value of conservation with the public. There is another point of critical miscommunication.

The phrase "conservation triage" is intended to highlight what some see as an apt analogy to medical triage. With medical triage, the largely empirical parameters (B, C, S) tend to be well known, and the normative parameter (V) is broadly agreed upon. There also tends to be a reliable understanding about both the degree to which resources will be inadequate throughout the duration of the crisis and the scope of harms during the crisis (e.g., in the day or two following a hurricane). None of those circumstances seem to apply so aptly to conservation triage.

If one were looking for a more appropriate analogy, the novel *Sophie's Choice* comes to mind. In the novel, World War II Nazis force a mother to decide which of her two children would be gassed and which would be sent to a children's work camp. If she refused to decide, the Nazis would kill both children. Sophie forsakes her daughter to the gas chamber, never sees her son again, and eventually commits suicide for the guilt she felt over making a decision for which there can be no rationale. It might be more strategic to replace the triage language with language more akin to an environmental Sophie's choice. Doing so might better emphasize that a sound decision about triage is likely impossible and to focus attention on the forces driving the decision, such as human population growth, corporate greed, government corruption, and consumption by the wealthiest among us.

Recap

We began this chapter with methods for conducting PVAs as a means of estimating $P[E_t]$. We discovered the following:

- *Accurate* estimates of $P[E_t]$ are virtually impossible for reasons represented by Table 6.1.
- While *useful* estimates of $P[E_t]$ are possible for a number of cases, they require far more data than are available for most species.
- While a universal MVP with a hoped-for level of precision does not exist, it is possible to incorporate PVAs into an adaptive management framework that would work for any species.

The ability to create useful estimates of $P[E_t]$ was facilitated by thinking of PVAs as a way to quantitatively connect threats, actions, and outcomes.

Expanding that framework to account for the cost of actions brought us to conservation triage and some of the most important ethical questions that arise from PVAs. Notice how considering those ethical issues depended on a reasonably technical understanding of the science of PVA, which are covered in the first part of the chapter. That dependency is worth noticing because many complicated ethical issues depend on a modicum of technical understanding.

Before we move on, please know that conservation triage is controversial and some professionals are very comfortable with the idea. Ultimately, conservation triage is a relatively new idea in a rapidly changing world. I expect views on conservation triage to evolve over the next few decades. In the online supplement I provide a list of papers that further explore this topic.

WHAT IS AN ENDANGERED SPECIES?

Having the ability to perform a PVA and estimate $P[E_t]$ is only part of the challenge. Those abilities beg the question: What levels of extinction risk are unacceptably high and should result in special protections and restoration efforts? While addressing that question requires significant ecological knowledge, it is ultimately an ethical question.

The question is also begged by key policies aimed at protecting endangered species. For example, the Convention on International Trade in Endangered Species (CITES) is a treaty that limits international trade in animal products for "all species threatened with extinction." The question raised by that treaty is, how threatened is threatened enough to warrant the protections of CITES? The US Endangered Species Act (ESA) offers special protections and restoration efforts to species that are "in danger of extinction." How in danger does a species need to be to warrant protections afforded by the ESA? Answers to those questions are debated and contested.

Acceptable Levels of $P[E_t]$

The two elements of $P[E_t]$—time and probability—can be represented on a graph with an axis for each element (Fig. 6.4). A PVA can be conducted to place a mark on this graph that describes the extinction risk of a population.[19] A PVA and the description of extinction risk that it produces are typically taken to be a largely or entirely scientific activity. After describing a population with a mark on this graph, one has to judge whether the risk is high enough to merit special protection and restoration.

One could make that judgment, as illustrated in Figure 6.4, by drawing a line across the graph to represent the boundary between "endangered"

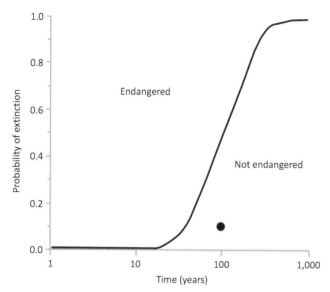

Figure 6.4 Assessing extinction risk. The most complete and precise assessment of extinction risk involves two dimensions—time and probability—as shown by this graph. Extinction risk is greatest in the upper left portion of the graph and least in the lower right portion. The dot represents a scientific description of extinction risk for a hypothetical population. The line represents a normative (ethical) judgment that categorizes some levels of extinction risk as unacceptably high and other levels as acceptably low.

and "not endangered." If the dot describing a population falls to the left of the boundary, then the population is endangered. Otherwise, not. Drawing such a line is not only an ethical judgment but also requires an understanding of probability theory, which only the most mathematically oriented among us have.

There has been a tendency for conservation professionals to gravitate toward $P[E_{100}] = 0.05$ as a boundary between endangered and not endangered (Doak et al. 2015; Shaffer 1981; Thompson 1991). No particularly good reason has been offered for this de facto ethical judgment.

Furthermore, that level of extinction risk ($P[E_{100}] = 0.05$ for individual species) corresponds to an extinction rate (frequency of extinctions among species) that is at least an order of magnitude, or more, higher than the natural background rate of extinction (Vucetich and Nelson 2018). Yet, solving the biodiversity crisis means reducing the extinction rate to some level much closer to the background rate. Thus, taking $P[E_{100}] = 0.05$ to be an acceptable level of extinction risk (for individual species) may be tantamount to describing and enabling the biodiversity crisis, rather than mitigating it.

Another concern about leaning too hard on this approach for judging which species deserve special protections is that it's not possible to get a sufficiently useful estimate of $P[E_t]$ for most species. If this approach were the only way to answer the question, "What is an endangered species?" then we'd have to live with those concerns. But it sure would be good to consider other possibilities.

Red List Criteria

The IUCN has a useful, sophisticated set of criteria for classifying species according to three levels of endangerment (Table 6.2). It includes multiple criterion types pertaining to P[E_t], including trends in abundance and range contraction. These criteria—taken as a whole—have been considered broadly useful.[20] But that usefulness is best judged by acknowledging that the architects of these criteria were explicit that their purpose is to provide a rank-ordered classification of extinction risk—not to make ethical judgments about what species deserve special protections and which do not.

Geographic Range

If there are shortcomings in judging what counts as an endangered species according to methods implied by Figure 6.4, is there an alternative approach? Yes, I think there is, and it begins by recognizing that the size of a species' geographic range is also an important predictor of extinction risk.[21] Furthermore, the essence of being an endangered species is *rarity*, which is a formal concept in ecology and rises from some combination of just two processes: diminishment of population density and loss of geographic range.[22]

Shrinking geographic range is a mechanism by which humans have elevated the extinction risk for many species, especially due to habitat loss and overexploitation. In particular, a majority of studied terrestrial vertebrates have been extirpated from 60% or more of their geographic ranges! Those species-specific losses in geographic range translate to large portions of the Earth's terrestrial surface having lost a significant portion of native species whose presence is required to confer their ecological value.* The scientists who discovered the extent of those losses wrote this, pertaining to the significance of the finding (Ceballos, Ehrlich, and Dirzo 2017):

> The strong focus on species extinctions, a critical aspect of the contemporary pulse of biological extinction, leads to a common misimpression that Earth's biota is not immediately threatened, just slowly entering an episode of major biodiversity loss. This view overlooks the current trends of population declines and extinctions. . . . [W]e show the extremely high degree of population decay in vertebrates, even in common "species of low concern." Dwindling population sizes and range shrinkages amount to a massive

---

*Important context is the first few pages of Chapter 1, especially Figure 1.1.

anthropogenic erosion of biodiversity and of the ecosystem services essential to civilization. This "biological annihilation" underlines the seriousness for humanity of Earth's ongoing sixth mass extinction event. (E6089)

The loss of biodiversity is a crisis, in part, because species are ecologically valuable to their native ecosystems. But a species cannot manifest its ecological value on portions of range no longer inhabited due to anthropogenic threats.

The take-home message is that the extent of range loss is an important basis for judging whether a species should be counted as endangered—that is, deserving of special protections and restoration.

Range and the US Endangered Species Act

Geographic range is also a key concept for the United States' legal definition of an endangered species.[23] According to the Endangered Species Act (ESA), an endangered species is any that is "in danger of extinction throughout all or a significant portion of its range." Any species meeting that criterion is entitled to the special protections of the ESA. There is good reason to think that that definition is meant to include any species extirpated from a significant portion of its historic range (Vucetich et al. 2024). That legal definition resonates with the problem the ESA is intended to solve—namely, the biodiversity crisis. The definition also begs a key ethical question, What portion of range loss is unacceptable?

Knowing the importance of that question, my colleagues and I surveyed a representative sample of Americans (Vucetich et al. 2021). Specifically, survey participants were presented with these statements:

> Extinction is a process that involves regional extinction at various places throughout a species' historic range. The geographic areas where a species lives is called their "range." Most mammal species have been driven to extinction from half or more of their historic range because of human activities.

Then, participants were asked:

> What percentage of historic habitat loss would be acceptable?
>   How much [what percentage] of a species' historic range should be lost before federal law steps in to protect a species?

Responses to those questions indicate that the majority of Americans believe that acceptable losses should be less than 30% of a species' historic range (Fig. 6.5).

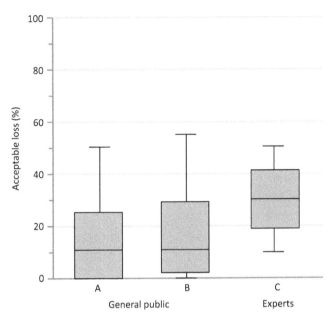

Figure 6.5 Judgments by the general public and experts on what constitutes an acceptable loss of geographic range. Boxes represent the 25th and 75th percentiles, horizontal lines within each box represent the median, and whiskers represent the 10th and 90th percentiles. The left column refers to the question, What percentage of historic habitat loss would be acceptable? The middle column refers to the question, How much [what percentage] of a species' historic range should be lost before federal law steps in to protect a species? See main text for other details.

Any judgment about acceptable loss depends on a synthesis of facts and values. As such, judgments from a representative sample of Americans could, in principle, be impaired by most Americans' lacking expertise pertaining to species endangerment. Being expertly informed about biological conservation may alter one's judgment about what counts as an acceptable loss.

This possibility is accounted for with another recent scientific finding. In particular, a representative sample of 438 conservation scholars were presented with this statement and question (Vucetich et al. 2021):

> In the past two centuries, the median terrestrial vertebrate species is thought to have lost approximately 60% of its historic range. Taking account of human needs, what would be an acceptable loss of historic range for most species (0 = no range loss and complete restoration; 60 = current median loss; 100 = complete extirpation)?

While experts tend to be accepting of larger losses, the distributions of responses for experts and the general public are importantly overlapping (Fig. 6.5). Three-quarters of the experts surveyed indicated that acceptable loss is 40% or less of a species' geographic range.

These results should not be the sole basis for answering the question of what is an endangered species. Nevertheless, if one wanted to be aligned with the typical conservation expert, one would judge an endangered species to be any that had lost 30% or more of its historic range. If one wanted to be aligned with what most Americans would accept, then one could also (coincidentally) select 30% loss of range as a boundary between endangered and not endangered.

Whatever you may think of this approach for judging endangerment, most professionals and most decision-makers do not stray far from the ideas associated with Figure 6.4 and boundaries on the order of $P[E_{100}] = 0.05$ (Doak et al. 2015). In any case, the salient point has been to show how one of the most important questions in population biology— What is an endangered species?—is an unresolved ethical concern and depends on the synthesis of knowledge from ecology, law, and the psychology of human values. This lack of resolution is not a concern for species such as the Bornean orangutan or the great tit. The former is clearly endangered and the latter is clearly not. However, Figure 1.1 and the text associated with that figure indicate that many species have lost enough geographic range as to render it uncertain whether they merit special protection or not.

HOPE, FATALISM, AND EXTINCTION

Habitat destruction, overexploitation, population decline, and range loss— they all lead to an extremely high rate of extinction that is the biodiversity crisis. The circumstances can seem hopeless, as though we're doomed to an impoverished future.

In response, some environmental leaders assert that being steadfastly hopeful about the future of our environment is essential to avoid unhealthy, chronic states of despair and the kind of fatalism that steals motivation to act for a better future.

I'm not so sure it's that simple. I'm not so sure that hope is the only thing that wards off such states of fatalism and despair. My concern is straightforward. Not everyone is hopeful about every important aspect of the future. And there are many instances in life where good reason leads one to think that certain important aspects of the future are likely to turn out poorly. For those people and in those cases, are fatalism and despair really the only alternative to being hopeful? I don't think so.

Some of the concern arises from the words themselves. "Hope," "fatalism," and "despair" are words whose meanings are remarkably nuanced and flexible.[24] Unhealthy states of despair need to be distinguished from manifestations of sadness that can be a healthy response to a sad situation. Hope also has subtly varied meanings. One might say, "I hope it doesn't rain tomorrow," even though the forecast says it's likely. In that case, one is really just saying, "I have no control over whether it's going to rain, and

I can see that it's likely to rain, but it's my desire that it doesn't rain." Alternatively, one might say, "I hope to get an A on the next exam," and mean, "I'm going to study hard, because if I do, there is a reasonable chance that I will get an A." Both instances of hope are reasonable uses of the word, but they convey different meanings. (To better internalize this point, draft a statement about the biodiversity crisis that uses the first meaning of "hope." Compare it to a statement about the biodiversity crisis that makes use of the second meaning of "hope.")

While mindful of the nuances of those words, I also appreciate that some people naturally and truly feel hopeful that massive improvements in our relationship with nature are just ahead. And I appreciate that hope inspires some of those people to work hard in contributing to those hoped-for improvements. That's all very fine.

But my worry is that this sense of hope might elude some people. For those people, is the only alternative to hope really fatalism? You might think this concern doesn't apply to you, because you feel hopeful. But what if some day in the future you come to the belief that humans will not abate the biodiversity crisis in time to prevent too many losses. What then?

Some moral philosophers offer another way to think about hope and fatalism. To see their view, begin by imagining a person who cares about the environment and knows that manifesting this care involves sacrifice. But this person also sees that too few others are making enough sacrifice to be of benefit to the environment. For that reason, this person gives up— they stop making sacrifices for the environment. One might say they've lost hope and become fatalistic about the future of the environment. But the moral philosopher might say this person is making their fulfillment in life too dependent on the actions of others over which they have no control. They do so, in this case, by judging their actions to be right and worthwhile only if those actions lead to a good outcome.

Now imagine another person who also cares for the environment and knows that manifesting this care involves sacrifice. But this person is different in having the strong belief that their fulfillment in life comes from discovering right principles to live by and striving to live by those principles—regardless of what others do and regardless of the outcome. This person would be saddened by the loss of biodiversity, but they are also a fulfilled and empowered person because they are motivated more by following right principles than by any expectation about how the future will turn out. This second person desires a positive outcome, and they will acknowledge sadness if the future doesn't turn out that way. But this person is ultimately motivated by doing the right thing—regardless of the outcome. For this person, hope doesn't really even enter the picture.

These two imaginary people represent two basic frameworks in moral reasoning, that is, consequentialism and deontology. You can read more about both frameworks and their relationship to the biodiversity crisis in the online supplement.

Philosophers are not the only ones to study hope and fatalism. Psychologists have attended to these topics, too. But most psychological inquiry has focused on hope in the context of dire medical prognoses (such as cancer) and their effect on patients' propensity to accept difficult treatments. Very little is known about the psychology of hope and fatalism in the context of climate change or the biodiversity crisis. For a toehold on what is known, see the online supplement.

Finally, let's remind ourselves of how hope and fatalism are related to the main topic of this chapter, which is the science of extinction. Environmental hope, environmental fatalism, and scientific knowledge about extinction—all three of those ideas pertain to one's expectations of the future. If the science of extinction leaves you feeling hopeful—very good. If that science leaves you with a heavy feeling, then perhaps this discussion about hope and the fulfillment that comes from principled action brings you some insight. At the very least, this discussion identifies an important connection between ecological science and moral philosophy.

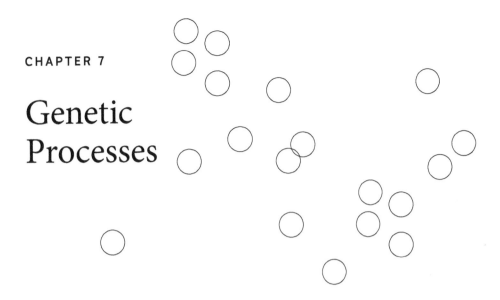

CHAPTER 7

Genetic Processes

I remember taking the image shown in Figure 7.1 from our observation plane on a cold, bright February morning as the wolves traveled across Lake Richie. The two wolves on the right are father and daughter. But they are also half-siblings, because they share the same mother. The third wolf—with the unusual posture and coarse-looking pelage—is their offspring. That wolf died sometime between the ages of 10 and 21 months. The other two wolves lived several more years, but never again reproduced. These are the last three individuals from the population of wolves that inhabited Isle Royale National Park, situated on the US side of Lake Superior, during the twentieth century. The wolf population on Isle Royale provides one of the best accounts of a small, isolated population of a wild vertebrate population going extinct due to inbreeding depression.

The case is important, in part, because for decades the population was assumed to be completely isolated from the mainland population, but it had in fact benefited from what geneticists refer to as *gene flow*. The population was also assumed to have escaped the detrimental effects of inbreeding, but it hadn't. The first real indication we observed for the toll inbreeding had been taking was a high incidence of bone deformities, like that shown in Figure 7.2. Those bones are the pelvis and sacral vertebrae of an Isle Royale wolf. The bones are supposed to be symmetrical, but clearly are not. When domestic dogs get deformities like this, it can cause pain and affect mobility.

I'll connect the wolves of Isle Royale to many of the ideas covered in this chapter. Then, I'll tell their story more completely at the end of the chapter.

Figure 7.1 Wolves from Isle Royale National Park.

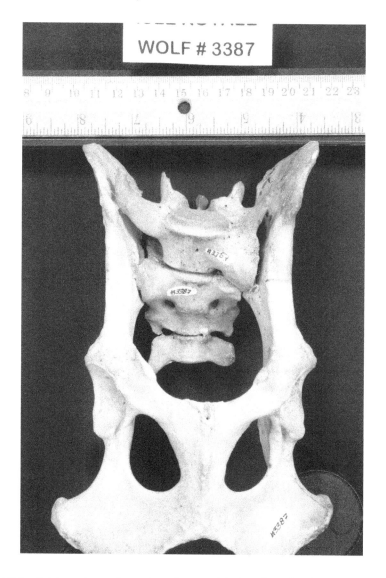

Figure 7.2 Pelvis and sacral vertebrae from an Isle Royale wolf. The asymmetries in the sacral vertebrae are likely a result of inbreeding. Credit: J. J. Henderson.

ROADMAP OF IDEAS

Understanding the relationship between the demographic and genetic dynamics of a population is a multi-leg journey. The path we'll follow is this series of questions:

- What is genetic diversity, and what processes affect genetic diversity?
- How does genetic diversity affect the fitness of individual organisms?
- How does genetic diversity affect the demography of entire populations?
- How much genetic diversity is needed to keep a population viable, and how does one know if a population has enough genetic diversity?
- What conservation actions can mitigate losses of genetic diversity to maintain or restore viability?

Although each question is interesting in its own right, you might be more drawn to those at the end of this list. If so, you'll soon find out how answers to the later questions depend on answers to the earlier questions. Let's go.

WHAT IS GENETIC DIVERSITY?

Think of an individual organism as a collection of *phenotypes*, that is, every measurable trait that might usefully describe that individual. Phenotypes include physical traits such as body size and eye color, behavioral traits such as temperament, and life history traits such as longevity. Phenotypes are importantly determined by an individual's genes. Each gene (or locus) in an individual's genome is represented by two copies, one of which is inherited from each parent.[1]

If the two copies of a gene are identical (e.g., if both copies are for blue eyes), then we say the individual is *homozygotic* at that locus. If the two copies are different (one for blue eyes and one for brown eyes), then we say the individual is *heterozygotic* at that locus. The proportion of heterozygotic loci in an individual's genome is a measure of genetic diversity. Every individual in a population has its own combinations of alleles[2] and history of parentage that led to those combinations. When all the genetic properties of all those individuals are scaled up to the population level, two interesting things happen.

First, we can recognize different ways of measuring and thinking about genetic diversity. For example, genetic diversity can be the property of an individual or a population. Thus, we can refer to the proportion of loci within an individual that are heterozygotic or to the proportion of heterozygotic individuals at a specific locus within a population. Other important measures of genetic diversity are focused on alleles, such as the number of different alleles at a specific locus in a population (such as the alleles that code for six different colors of eyes in humans). Another allelic-focused metric of diversity is allele frequency. For example, the allele frequency for a gene associated with late fall migration had been about 0.13 in a population of pink salmon. From 1980 to 2010 the frequency of that allele had dropped to 0.04—almost certainly driven by climate change (Kovach, Gharrett, and Tallmon 2012).

The second interesting thing that occurs as we broaden our focus from individuals to populations is that those aspects of genetic diversity change over time and differ between populations. These changes and differences are the subject of a scientific field known as *population genetics*. More specifically, changes in allele frequency over time is called evolution. When population genetics is studied and applied for the purpose of conservation, the subject is called *conservation genetics*. From the perspective of conservation, genetic diversity may be considered valuable for its own sake, but it is also demonstrably valuable to the demographic viability of a population.

WHAT PROCESSES AFFECT GENETIC DIVERSITY?

Changes in genetic diversity over time result from the combined influence of several processes. We'll examine each of these processes in the next several sections. The first sections may be a brief review for you, but we will quickly get into new material.

Mutations

Mutations are the ultimate source of all new genetic diversity. Most mutations are not especially consequential (e.g., have no effect on phenotype) or are so severely detrimental that they never spread in a population because they tend to kill individuals at a young age. But some mutations are adaptive, and those that are tend to persist in a viable population. Some mutations are maladaptive, but not so maladaptive as to be eliminated from the population (sickle cell anemia is an example).

Natural Selection

Natural selection can change genetic diversity in various ways, depending on the kind of selection. There are three main kinds of selection: balancing, directional, and disruptive.

Balancing selection tends to maintain genetic diversity by preventing large increases or decreases in diversity over time. Balancing selection is easiest to understand through an example such as body size. Think about moose. Individual males that are too small will be disadvantaged; for example, they won't be able to compete with other males for access to females. Individuals that are too large will also be disadvantaged; they may have nutritional requirements that are not easily met, which can, for example, preclude their immune systems from getting enough energy. Alleles leading to more extreme body sizes will tend to be eliminated from the population by balancing selection. Balancing selection is also referred to as *stabilizing selection.*

Directional selection is also easier to first grasp through an example. In Kluane National Park (southwestern Yukon, Canada), springs have been warming, and this change favors red squirrels who mate earlier in the spring. Over the past few decades, many squirrels have begun to consistently mate earlier in the season. While much of that shift in timing was due to phenotypic plasticity,[3] some of the shift was due to directional selection (Réale et al. 2003). Directional selection is expected to continue making alleles associated with late reproduction rare or extinct and alleles associated with early reproduction more common.

In the interest of keeping focus on the big picture, and because I've conveyed the most salient points by discussing directional and balancing selection, I won't describe disruptive selection. But you can easily learn about disruptive selection with a quick internet search.

Nonrandom Mating

The most basic elements of genetic theory assume that organisms select mates at random. When certain mate pairings are more likely than others, then mating is nonrandom. The results of nonrandom mating can affect gene frequencies over space and time, and more complex theory is required to understand the dynamics.

> Space and distance are important causes of nonrandom mating, which has significant consequences for genetic diversity, as illustrated in Figure 7.3 and further explained later in this chapter.

Space is an extremely important cause of nonrandom mating. Very often, individuals living near one another are more likely to mate than those living far from one another. As a result, closely situated populations (of the same species) tend to have more similar genetic diversity than populations that are farther from one another. The nonrandom mating associated with space is also associated with a species' capacity to disperse:

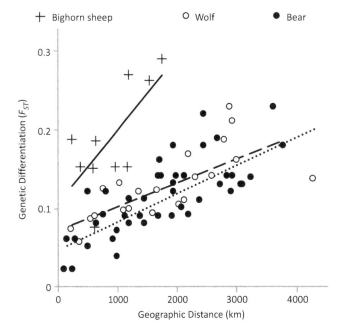

Figure 7.3 Genetic differentiation as a function of geographic distance for three species occupying large geographic ranges. Differences in genetic diversity between populations (quantified by a metric known as F_{ST}) tend to increase as the distance between the populations increases. Each symbol is the differentiation between two populations. Source: Forbes and Hogg (1999).

species characterized by shorter dispersal distances tend to have greater divergences in genetic diversity over space (Fig. 7.3). This dispersal ability is sometimes an innate feature of a species, while on other occasions it is critically influenced by habitat fragmentation.

Assortative (or disassortative) mating occurs when individuals of similar (or dissimilar) genetic-based phenotypes are more likely to mate with each other (Jiang, Bolnick, and Kirkpatrick 2013). For example, taller humans are more likely to reproduce with each other (assortative). And white-throated sparrows have a strong tendency to select mates with different-colored stripes over the eye (disassortative) (Hedrick, Tuttle, and Gonser 2018). While the consequences of these mating patterns are complex and varied, there is a tendency for assortative mating to increase homozygosity and be associated with disruptive selection and for disassortative mating to increase heterozygosity and be associated with stabilizing selection.

Sexual selection is another important kind of nonrandom mating and occurs when individuals of the same sex compete for mates (Clutton-Brock 2007). That competition can result in some sex-specific phenotypes having a greater chance of being reproductively successful. The antlers of males for many cervid species are an example of sexual selection.

Inbreeding

After breezing through those first three processes, we'll slow down to focus on the finer details of inbreeding. Inbreeding is best thought of as a

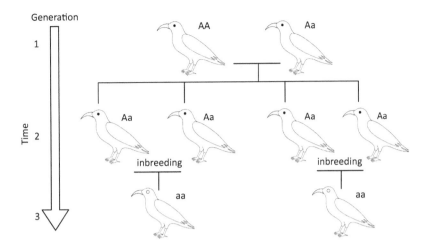

Figure 7.4 Hypothetical pedigree illustrating inbreeding in the form of reproduction between siblings. The middle row shows four siblings, among which two matings occur. All four siblings were heterozygotic (Aa), and their mating resulted in homozygotic (aa) offspring. Furthermore, the frequency of "a" was 0.25 in generation 1 and increased to 1.00 by generation 3. Allelic diversity had been two (a and A) and declined to one (a). The relationships in this pedigree did not have to result in these temporal fluctuations in allelic frequency or diversity. Rather, they are one of several possible outcomes. It is also possible, for example, for generation 3 to have been comprised of two heterozygotic individuals. The genetic outcomes associated with a pedigree have an important random (stochastic) component. This random component is why *F*, the inbreeding coefficient, can be defined in terms of probabilities (see main text for details). This pedigree also illustrates how inbreeding is problematic because it tends toward increased expression of deleterious, recessive alleles ("a" in this example). Such a problem exists for red-billed choughs of Scotland, who occasionally hatch nestlings that are blind and never fledge (main text for details). In this diagram, birds exhibiting the blind phenotype have white eyes.

constellation of related ideas all referring to various kinds of nonrandom mating. The simplest understanding of inbreeding is what occurs when close relatives living in the same population mate with each other. More precisely, inbreeding occurs when mated individuals are more closely related to one another than would occur if two individuals had been selected at random from a large population.[4] In any case, the closer the relatives, the more intense the inbreeding. To further understand inbreeding and its consequences consider a pedigree, which is like a flowchart showing the history of mating and parentage in a population (Fig. 7.4). It turns out that if you know everyone's parents, then you can also determine how everyone is related to everyone else in the population.

In a pedigree, the passage of time is represented by moving from the top to the bottom of the pedigree. Pedigrees can also be presented to highlight individuals' phenotypes or genotypes of interest. Highlighting genotypes in a pedigree also emphasizes how the genetic diversity of a population changes as old individuals die and new individuals are born. For example, the pedigree in Figure 7.4 happens to represent an instance of very intense inbreeding—that is, mating between very close relatives, in this case, siblings.

The pedigree in Figure 7.4 illustrates this general process: when close relatives mate, there is a greater chance that an individual is born with two copies of the same allele. For that previous sentence to make sense, think of the lower case "a" in Figure 7.4 as literally being the same allele that's been copied over and over. The more it gets copied, the more likely it is that two copies will show up in the same individual. From this pedigree follows a simple and important lesson: inbreeding is associated with a tendency to lose genetic diversity. Soon, we'll also see that the more intense the inbreeding, the faster the rate at which diversity is lost.

There's more. Recall from an introductory biology course that an allele's influence on a phenotype can sometimes be described as dominant or recessive. The phenotype associated with a recessive allele (e.g., blue eyes in humans) is realized only if both alleles (from the mother and the father) are of the recessive type. Otherwise, the phenotype will be that which is associated with the dominant allele (e.g., brown eyes).[5]

Inbreeding is mating with close relatives and is concerning because it can lead to the expression of rare, deleterious, recessive alleles.

Deleterious alleles are associated with maladaptive phenotypes, and they are often recessive. The deleteriousness of an allele can be so severe that individuals with two copies of the allele are all but certain to die before they themselves can reproduce. These are lethal alleles, and they also tend to be recessive.

The effect of lethal alleles is illustrated by red-billed choughs of Scotland. They are highly endangered members of the Corvidae family of birds and distinguished by a long, curved, red beak. Red-billed choughs are declining throughout Europe. The Scottish population is isolated on the islands of Islay and Colonsay off the southwest coast of Scotland and consists of fewer than 50 pairs. As such, the birds of this population are vulnerable to inbreeding. Furthermore, researchers noticed, beginning in 1998, that some nestlings were born blind and as a result never survived to fledge. By studying patterns of inheritance among and between families, researchers deduced that the blindness phenotype is also recessive (Trask et al. 2016).

Another example of lethal recessive alleles that is problematic for conservation is an allele that leads to extreme dwarfism (chondrodystrophy) in California condors. Also of concern are Florida panthers who exhibit a high incidence of cryptorchidism, in which a testis fails to descend from the abdominal cavity, resulting in an impaired ability to reproduce. The genetics of cryptorchidism are not fully worked out, and multiple genes are probably involved. But recessive genes are likely giving rise to this trait.

The best evidence to date is that most mammal populations have about 12 lethal equivalents. The meaning of that statement requires a little unpacking. First, an allele is said to be lethal if it causes the death of an individual that is homozygotic with that allele for that gene. If an individual had one such allele, we'd say they have one lethal equivalent. But there's more. An allele could confer a 50% chance of death if an individual were homozygotic with that allele for a gene. This allele represents ½ a lethal equivalent. An individual with 2 such alleles for 2 different genes, would have 1 lethal equivalent (½ + ½ = 1). If an individual had 3 alleles for 3 different genes, and if each of those alleles conferred an 80% chance of death, then we'd say that individual has 2.4 lethal equivalents (0.8 + 0.8 + 0.8 = 2.4). Returning to the opening sentence of this paragraph, an individual with 12 lethal equivalents could have 12 lethal alleles or 24 alleles that would lead to a 50% chance of death if the individual were homozygotic for any of those traits, or 48 alleles each leading to a 25% chance of death, or any number of detrimental alleles whose lethality "sum" to 12.

Finally, not all recessive alleles are detrimental. For example, long-haired cats and red-haired humans get those traits from recessive genes. Furthermore, not all detrimental alleles are recessive. Some are dominant, such as the allele that gives rise to a maladaptive egg-laying behavior in the endangered black robins of the Chatham Islands off the coast of New Zealand.

By 1980, black robins had been reduced to just five birds, which included only one mature female. During the 1980s, intense conservation efforts included monitoring the nest success of color-banded individuals and moving some eggs from robin nests to be raised by foster mothers of a related species (tomtits). Doing so caused the robins to lay more eggs, resulting in more rapid population growth.

During this period of intense conservation, it was also noticed that some robins laid their eggs on the rim of the nest and that these eggs tended not to hatch. To increase the population's growth rate, the conservationists made a practice of relocating those eggs toward the center of the nest. Doing so, however, led to an increase in the frequency of the allele that caused this maladaptive behavior. By 1989, 50% of the females exhibited the behavior. Upon realizing what was happening in 1989, conservationists stopped moving rim-laid eggs to the center of nests. By 2011, the frequency of rim-laying females was down to 9% (Massaro et al. 2013; Crew 2019). Pedigree analysis showed that the deleterious phenotype was caused by a dominant allele.[6]

Inbreeding Coefficients

Inbreeding is not a yes-or-no process. Rather, it comes in degrees of severity that are quantified by the inbreeding coefficient (represented by the symbol F), which is a number ranging from zero to one, where zero represents no inbreeding and one is the most intense inbreeding. F is usefully defined as the probability that an individual will inherit two alleles that are identical copies of each other that had appeared in an ancestor's genome.[7]

When defined in this way, an individual's F depends on the relationships in a pedigree. For example, individuals whose parents are full siblings have an F of 0.25 (meaning that such individuals have a 25% probability of inheriting two identical copies of an allele at any particular locus). Individuals have an F of 0.125 if, for example, their parents are half-siblings or if their parents are an aunt and a nephew. For a final example, individuals whose parents are first cousins have an F of 0.0625. For context, the two adult wolves in Figure 7.1 have inbreeding coefficients of 0.125 and 0.31, and the juvenile in that image has an inbreeding coefficient of 0.44.

The letter F is also used to represent the average inbreeding coefficient among individuals in a population. At the population level, inbreeding can accumulate over time (Fig. 7.5). Soon we will see that inbreeding accumulates more quickly in small populations, and we'll explore what's known about how large a population needs to be to avoid the detrimental consequences of inbreeding.

Genetic Drift

Genetic drift is a genuinely abstract idea, but the work required to understanding it is well worth the effort, because this process lies at the heart of much that concerns the health of animal populations. To understand drift, you will have to temporarily suspend your beliefs about what a population even is. You'll benefit from a very loose imagination. And you'll have to trust that I know where we're headed when I say that a population is like a jar of marbles.

Further, suppose that we do not even care that a population is comprised of individual organisms. Rather, what we care about are all the gametes

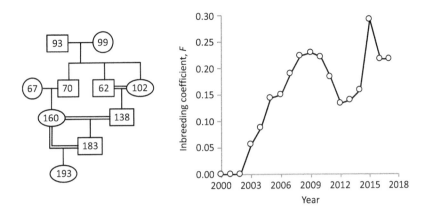

Figure 7.5 Partial pedigree from the wolf population in Isle Royale National Park (left panel), showing instances of breeding between close relatives. Individuals connected by double lines represent especially close matings. From a more complete pedigree, the average *F* of each wolf alive each year was calculated. The temporal fluctuations in that average *F* are shown in the right panel. *F* increases for a number of years, as expected. The decline in *F* occurs due to an important complexity that I'll explain later in the chapter. For context, the two adult wolves pictured at the beginning of this chapter are wolves 183 and 193 in the pedigree. Source: Hedrick et al. (2014, 2017).

(eggs and sperm) in a population—that's all. Recall that each individual in a population has many gametes and that each gamete carries one allele per locus. For the sake of simplicity, we'll track just one locus. Each marble in our jar represents a gamete, one of the many, many gametes in the population (Fig. 7.6).

Let's keep supposing. This population has three kinds of alleles, represented by the marbles' colors. Let's say black, gray, and white. The frequency of the colors in the jar is the allele frequency. Those frequencies (and how they change over time) are what we care most about. The gametes/alleles that make up the population's next generation are determined by drawing marbles from the jar at random. If the population is to have 500 individuals, then we'll pull 1,000 marbles (because each new individual gets a gamete/allele from each of two parents).

If the frequency of colors is 0.25 (black), 0.50 (gray), 0.25 (white), then you'd expect to draw 250 black, 500 gray and 250 white marbles. Due to luck (stochasticity), you would not be surprised to have drawn 271 black, 487 gray, and 242 white marbles. If so, then the allele frequencies for this next generation have changed to 0.271, 0.487, and 0.242. If you draw another 1,000 marbles for the next (third) generation, then you'd expect to pull marbles at that new frequency, but again, you wouldn't be surprised if the new frequencies drifted a little bit to maybe, 0.255, 0.478 and 0.267. This

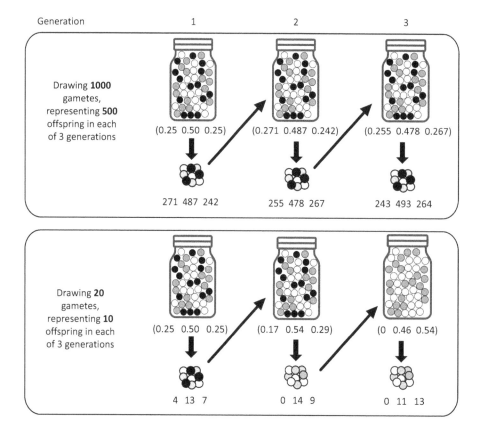

Figure 7.6 An analogy for genetic drift. It can be understood by supposing that the transmission of gametes and alleles from one generation to the next is similar to selecting a fixed number of colored marbles from a jar containing many, many marbles. By chance, the frequencies of different colored marbles selected for the next generation will differ from their frequency in the jar. The smaller the population (i.e., the fewer marbles that are selected), the more likely one will observe a larger shift in frequencies of alleles (different colored marbles) from one generation to the next. See text for details.

evolution—this temporal fluctuation in allele frequency—is said to be due to genetic drift. Granted, the fluctuations are small. But, in a moment we'll see that the smallness of those fluctuations is exactly the point, given that this imaginary population is relatively large (i.e., 500 individuals).

Now rerun this same weird thought experiment. This time suppose the population is small, say only 12 individuals, which means we'll be drawing 24 marbles (gametes/alleles) for each new generation. You'd expect to draw 6 black, 12 gray, and 6 white marbles, which corresponds to the starting allele frequencies from the previous thought experiment—that is, 0.25, 0.50, and 0.25. In spite of that expectation, you wouldn't be surprised if—by luck (stochasticity)—you drew say 4, 13, and 7 marbles. That se-

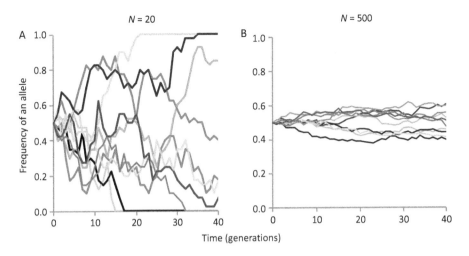

Figure 7.7 Fluctuations in allele frequency over time for populations of different size. If the genetic drift process described in Figure 7.6 is allowed to play out for many generations, then the frequency of an allele fluctuates randomly over time, as shown here. Each line in each panel represents a different population exposed to the identical process of randomly selecting alleles (gametes) for the next generation. The only difference between the two graphs is the size of the population (*N*). The rate at which frequencies change (diverge from their original value) is greater in populations that are smaller. Eventually the allele frequencies drift to either zero or one.

lection means the new allele frequencies are 0.17 (= 4/24), 0.54 (= 13/24), and 0.29 (= 7/24). Do it again, and you draw 0, 14, and 9 marbles. Do it again and you might draw 0, 20, and 0 marbles. In just three generations the allele frequencies shift so greatly that two of the alleles went extinct—just by luck. The point is: the allele frequencies changed so wildly because the population was small.

That thought experiment illustrates an important aspect of the process of genetic drift. But it would also be good to see how that process plays out over longer periods and for many different populations. To do so, one can write a simple computer program, the result of which is shown in Figure 7.7. While each population's trajectory of allele frequencies is unique, the general pattern is easy to spot. When *N* is smaller the allele frequencies drift from their original values more rapidly. In other words, the rate of genetic drift increases as population size decreases.

When *N* is small, allele frequency drifts, often enough, all the way to zero (indicating the loss of the allele) or one (indicating the loss of all the other alleles in the population). Lost alleles won't be restored to the population unless there is a chance mutation or unless an immigrant from another population restores the lost allele. The former is not especially likely, and the latter is a subject that we cover later in the chapter.

Now it's time to ask, just how representative of a real population is that jar of marbles? The answer is more representative than one might initially think. If every individual in a population contributes equally to the next generation and if abundance doesn't fluctuate over time, then the jar-of-marbles analogy is a pretty good representation of how genetic drift works in a real population. Furthermore, there are ways to understand how genetic drift would change if population abundance did fluctuate or if individuals did contribute unequally to the next generation. Again, this is a subject covered later in this chapter.

Inbreeding, Drift, and Population Size

Now is a good time to make some comparisons and contrasts between genetic drift and inbreeding:

· Genetic drift is primarily about the loss of alleles due to random processes associated with sampling a finite number of alleles to make each new generation. That loss of allelic diversity will result in a decrease of heterozygosity.[8]
· Inbreeding is primarily about the relatedness of mated pairs and the resultant loss of heterozygosity. Allelic diversity will also be lost through mating between close relatives (e.g., Fig. 7.4).
· Both processes occur more intensely as population size decreases.

But how small is small? The answer provided in the next paragraph is a little telegraphic (i.e., at risk of being too succinct). Consequently, you might take one of three options with this upcoming paragraph. First, you might understand every step perfectly. Second, you might find a population genetics textbook for a more thorough explanation. Third, you might, more modestly, appreciate that there is a mathematical logic (gestured to in the next paragraph) and know that this logic leads to some useful equations and insightful principles that follow from the equations. My hope for you—right here and now—is that third outcome. The others are very achievable, and the degree to which you pursue those paths is between you and your instructor.

An important way to be homozygotic at a locus is for two identical alleles to be inherited from related parents (Fig. 7.4). Because a population of diploid individuals has $2N$ alleles at any particular locus, the probability of such an inheritance for any given mating is $1/2N$.* Because matings

---

* Diploid organisms have paired chromosomes, opposed to haploid organisms which have only one set of chromosomes. Most multicellular organisms are diploid.

accumulate over time in a population, so too does the probability of multiple such inheritances. That circumstance leads to this equation:

$$F_t = 1 - (1 - 1/2N)^t, \tag{7.1}$$

As the rate of inbreeding increases, genetic diversity is lost at a faster rate.

which describes the population-level inbreeding coefficient at generation t. F_t is also the proportion of individuals that are expected to be homozygotic (due to having inherited identical copies of an allele). The remaining individuals in a population would be heterozygotic. As such, $F_t = 1 - H_t$, where H_t is the expected heterozygosity in generation t.

That simple relationship means equation 7.1 can be rewritten in terms of H_t:

$$H_t = H_0(1 - 1/2N)^t. \tag{7.2}$$

Rewriting the equation in this way is important for a special reason. First, F is difficult to measure for a wild population because the information needed to make a pedigree is difficult to gather. However, it has become easier (in recent decades) to estimate the frequency of heterozygotes in a wild population by laboratory analysis of samples containing the DNA from the populations' individuals. The ability to make lab-based estimates of H is important for connecting the theory of inbreeding to genetic properties of real populations.

Use Excel and Equation 7.1 to confirm the following two results. First, F_t increases over time, no matter the value of N. Second, the rate of increase in F_t is greater for smaller populations.

For example, if you do the work described in the gray box, you'll have reason to think that larger populations should tend to have more genetic diversity. Sure enough, larger populations do tend to have higher values of lab-based estimates of H (Fig. 7.8).

The preceding account focuses on the relationship between population size and inbreeding (F). Most precisely, and compactly, the rate at which inbreeding accumulates from one generation to the next is $1/2N$. What about the relationship between population size and genetic drift? Figure 7.7

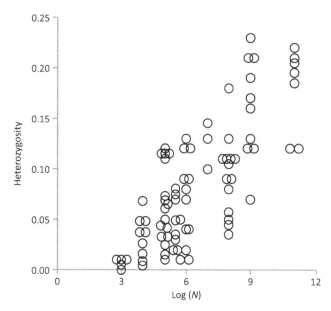

Fig 7.8 Heterozygosity in relation to population size. As population size increases (along the x axis), there is an important tendency to see an increase in the frequency of heterozygotes, as estimated from the laboratory analysis of DNA collected from individuals belonging to the population. Source: Soulé (1976), reprinted in Frankham (1996). The symbols represent different populations of various species.

already gives the impression that the rate of drift increases with decreasing N.

In your mind, draw a vertical line through each graph in Figure 7.7 at, let's say, $t = 20$. Use that vertical line to help focus on the idea that each population in those graphs has a different allele frequency at $t = 20$.

You could calculate the variance in allele frequencies at $t = 20$ and find that the variance is larger for the graph representing smaller populations. Furthermore, if you move that vertical line (with your mind) to the right, you'll see that the variance increases. In the 1930s, a very bright population geneticist presented the mathematical reasoning required to conclude that the influence of genetic drift on that variance in allele frequency depends on time and abundance in this way:

$$\mathrm{Var} \approx (1 - \exp(-t/2N))/4, \tag{7.3}$$

where t refers to time, measured in generations, and N is population size (Wright 1931). This particular equation holds if the initial allele frequency is 0.5. Other equations have been derived to predict how much time is expected to pass before an allele frequency becomes fixed at zero or one. The mathematical reasoning required to derive such equations is more involved than we need to get into here. If you want to know more, you can consult a textbook devoted to population genetics. One point of my showing this equation (without deriving it) is to show that the consequences of genetic drift can be predicted from mathematical formulas. A second point of doing so is to show that the rate at which genetic drift proceeds is related to the expression $1/2N$. In other words, drift progresses more quickly in smaller populations.

> As population size decreases, the rate of genetic drift increases. More specifically, the rate of drift is related to the expression $1/2N$.

Bathtub Metaphor

These processes—genetic drift, inbreeding, selection, mutation, and non-random mating—all operate simultaneously to determine the amount of genetic diversity in a population. To better understand how, think of a population's genetic diversity as water in a bathtub. Genetic diversity is added by some processes (at the faucet) and lost due to other processes (down the drain). The amount of diversity (water in the tub) depends on the *rates* at which diversity is added and lost. This is why it is so important to appreciate that the *rates* of inbreeding and drift increase as N decreases. Soon, we'll see how large N needs to be to avoid losing genetic diversity too quickly.

We'll also see how large a population needs to be so that the rate of drift is weak enough so as not to overpower selection's tendency to remove deleterious alleles from a population. And by the end of this chapter, we'll see how gene flow—the movement of genetic material between populations via dispersing individuals—can also offset losses in genetic diversity that occur due to drift and inbreeding. Before covering those topics, we'll learn a little more about inbreeding depression, which is the bad stuff that tends to happens to organisms that are too inbred.

INBREEDING DEPRESSION

The theoretical foundations of population genetics—ideas such as genetic drift and inbreeding coefficients—were set in the 1930s. The next big step in population genetics was to begin testing those theoretical ideas with laboratory experiments. Many such experiments were conducted on fruit flies (*Drosophila*) because they are easily raised in a lab and have short generation times. Those experiments began in earnest during the 1940s. During the 1980s, significant knowledge was being gained by studying warm-blooded animals that had been raised in zoos. From the 1990s onward, more and more insight has been gained by studying wild populations.

Zoo Populations

Before focusing on wild populations, let's learn something important from zoo-raised animals. More specifically, let's talk about pygmy hippopotami. They are endangered, native to the wet forests and swamps of west Africa, and their abundance in the wild has declined to probably fewer than 2,500. Another 350 or so pygmy hippos live in zoos throughout the world. For many decades, they have been bred in captivity, and zookeepers have tended to control and keep track of which hippos mate with each other. Sometimes one particular hippo is temporarily relocated for the purpose of mating with another particular hippo. Zookeepers keep track of such information in a *studbook*. By knowing who the parents of each individual are, it is possible to build a pedigree and calculate F for individuals in the population. These studbooks contain other basic information such as the dates of birth and death for individuals.

With that added information, it has been possible to create a graph showing how an individual's inbreeding coefficient is related to the probability that they will die as a juvenile (Fig. 7.9, panel A). Panel A of Figure 7.9 makes clear that inbreeding (increased F) is bad for individual pygmy hippos. Is that result particular to pygmy hippos or representative of most species? The only way to know is to examine the relationship in panel A of Figure 7.9 for a bunch of species for which studbooks have been kept. That line of thinking was pursued by Kathy Ralls and Jonathon Ballou in the 1980s. They identified 44 species of zoo mammals for which the necessary information existed. They made a graph like that in panel A for each species. On each graph they observed two values. The first value is called the *outbred mortality*, which is the expected mortality rate when $F = 0$. The second value is called *inbred mortality*, which is the expected mortality for the largest observed value of F. For the pygmy hippo, these mortality rates are approximately 0.25 and 0.55, respectively (Fig. 7.9, Panel A). That increase in mortality with increasing F is a specific instance of inbreeding depression.

Ralls and Ballou (1983) calculated 44 pairs of outbred and inbred mortality, with each pair of values representing a different species. Next, they thought of a clever way to graph all those pairs of values as points on a single graph to get an idea for the severity of inbreeding depression while taking account of many species. This new graph had inbred mortality on the x axis, outbred mortality on the y axis, and a diagonal line to serve as a reference, representing points on the graph where inbred and outbred mortality are equal (Fig. 7.9, panel B). Species whose studbook data placed it below the reference line are species for which inbreeding was detrimental. Conversely, species whose studbook data placed it above that reference

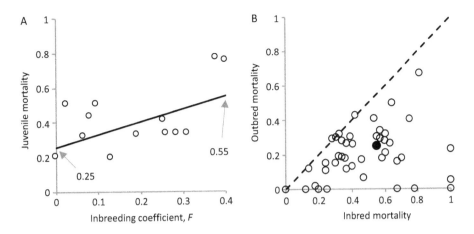

Figure 7.9 The assessment of inbreeding depression in zoo populations. The left panel shows the relationship between the risk of juvenile mortality and inbreeding coefficient for pygmy hippopotami raised in zoos. The solid line is a trend line, and the numbers highlight expected mortality rate for the outbred individuals (*y* intercept) and the most inbred individuals. The right panel shows the relationship between mortality rate of the most inbred individuals (*x* axis) and outbred individuals (*y* axis) for 44 different species of mammals that had been raised in captivity. The dotted line is a reference, indicating where inbred and outbred mortality are identical. Observations below this reference line indicate presence of inbreeding depression. The black symbol in B represents pygmy hippos (the point [0.55, 0.25]). Sources: Ralls and Ballou (1983), Frankham, Ballou, and Briscoe (2002).

line are species for which inbreeding had a positive effect on juvenile mortality. Species that were close to the line would be species for which inbreeding had little effect on juvenile mortality.[9] When Ralls and Ballou assembled their graph, they found that most species had been detrimentally impacted by inbreeding.

Some Skepticism

An active debate began in the 1980s between two groups of conservation scientists. One group tended to be less knowledgeable about population genetics and doubted its importance for conservation. The other group tended to be more knowledgeable about population genetics and warned that conservation goals failing to account for population genetics would fall short. The debate was fueled, in part, by skeptical thoughts such as:

> While inbreeding depression impacts certain vital rates in laboratories and zoos, that impact may be greatly reduced in wild environments for any of several reasons. For example, suppose that inbreeding depression impacts juvenile survival in the wild. Fine, but that impact is liable to be compensated for by increased vital rates in other individuals. In other words, density-dependent

processes are liable to absorb any detrimental influence of inbreeding depression. Furthermore, the severity of inbreeding depression observed in lab and zoo populations may occur only because some of those individuals have very high values of F. In the wild, F may not tend to get high enough to result in significant inbreeding depression.

Skepticism about the relative importance of population genetics has subsided greatly—though not entirely. For example, in the 2010s, I alerted park managers that the wolves of Isle Royale were at great risk of going extinct due to inbreeding depression. I had been accused of calling a false alarm ("crying wolf," actually) by a fellow scientist who had the ear of decision-makers. Nevertheless, the focus today tends to be not on *whether* population genetics should be taken into account, but rather more precisely *how* to do so.

Wild Animals

Since the 1990s, a great deal of evidence about inbreeding depression in the wild has become available. Here, we'll examine a case that provides a feel for both the effort required to study inbreeding depression in the wild and the kind of results that have been obtained.

The population that we'll consider is located on Mandarte Island, off the coast of Vancouver Island (British Columbia, Canada). While it is a proper island, Mandarte leaves the impression more of a low flat rock peeking out of the ocean. The island is loosely draped with low-lying vegetation and is home to a population of song sparrows. Every song sparrow on the island has been monitored (mainly through color banding) since 1975. Researchers have been monitoring everything about this population that they can—overwinter survival rates, nesting success, territory size, and more. They know which pairs of birds have nested together, and they've banded their offspring. They've drawn blood from many of these birds to analyze blood chemistry and DNA. From those efforts, they've confirmed the parentage for many of these birds.

With all that work, they've also been able to build a pedigree and calculate F for many of the population's individuals. In doing so, they found values of F ranging from 0 to approximately 0.25, indicating that some of the birds were considerably more inbred than others. For many of these birds, they also have information about traits that are likely to contribute to the birds' fitness.

One of the investigated traits was a particular kind of immune response called *cell-mediated immune response*. Assessing this trait involves pricking the skin of a bird with phytohaemagglutinin on the patagium (a fold of skin on the wing's leading edge). The prick results in temporary inflammation due to having stimulated immune cells at the site of the prick. The intensity of inflammation—measured about 12 hours after injection—is an index of the robustness for a key part of a bird's immune system. Researchers performed these skin pricks on a number of birds. As you might have guessed, sparrows with the lowest F tended to have the most robust immune response (Fig. 7.10, panel A).

Another trait that researchers investigated was song repertoire. Like the feathers of a male peacock or antlers of a male deer, the songs of a male

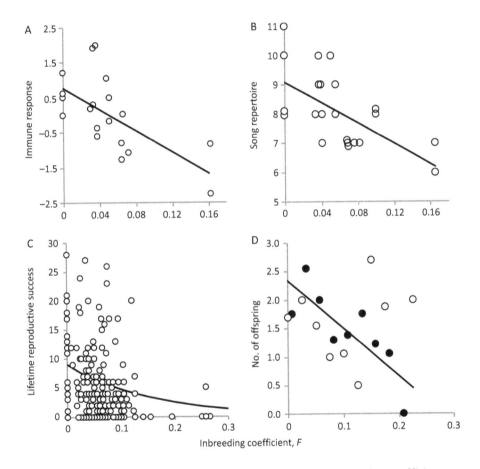

Figure 7.10 Fitness-related phenotypes in relationship to inbreeding coefficient (F) for a population of song sparrows living on Mandarte Island. The trendline for panel C predicts an expected lifetime reproductive success of 9 offspring for $F = 0$, 4 offspring for $F = 0.125$, and <1.5 offspring for $F > 0.3$. In panel D, filled symbols are data collected during cool springs, and the solid line is the trend line for those data. The open symbols in panel D are data collected during warm springs; for those data there is no significant trend between F and number of offspring. Sources: Reid et al. (2005), Figure 13.20 in Allendorf, Luikart, and Aitken (2012), and Figure 7.4 in Keller and Marr (2006).

songbird are secondary sexual "ornaments." These ornaments are used, in part, to attract female mates and compete with other males over access to female mates. Exaggerated ornaments are thought to be advantageous to an individual's fitness. With respect to songbird songs, exaggeration means knowing and singing a larger repertoire of songs. The number of songs that a song sparrow knows varies among individuals, commonly from as few as five to as many as ten songs. Each song is only a few seconds long, but if a song sparrow is really serious, he will sing each song repeatedly before moving to his next song. An entire repertoire of songs can take up to a half hour to sing.

The question is, how is song repertoire related to inbreeding? Again, you might have guessed that sparrows with the lowest F tended to have larger repertoires of songs (Fig. 7.10, panel B). Furthermore, song sparrows with larger repertoires tend to have more offspring and grand offspring.

Those findings—about songs and immune responses—are impressive and important because those traits are likely associated with an individual's fitness. Nevertheless, what would be really valuable is to understand how inbreeding affects traits that are even more closely related to fitness and population dynamics—traits such as reproduction and survival.

The sparrow researchers of Mandarte Island have made those assessments, as well. More specifically, they found that sparrows with the lowest values of F tended to have the greatest lifetime reproductive success—a trait that takes account of both an individual's survival and reproduction. The most inbred birds had, on average, fewer than one-third of the offspring of outbred birds (Fig. 7.10, panel C).

Finally, these researchers observed the relationship between a father's F and the number of offspring raised each spring over a number of years. On first appearances, the relationship was not especially clear (Fig. 7.10, panel D). But clarity comes from knowing that the song sparrows of Mandarte Island, living as they do near the Pacific Ocean at 48°N, are affected by spring temperature, which varies considerably from year to year. These sparrows tend to do better during warmer springs. Taking account of that information, the researchers found that inbred fathers tended to raise about as many offspring as non-inbred fathers during years with warmer springs. But, during years with colder springs, the inbred fathers were at a disadvantage when it came to raising offspring. The important lesson of that finding is that stressors in the environment—climate, predation, disease— are likely to magnify the severity of inbreeding depression.

The sparrows of Mandarte Island are just one of many wild populations for which inbreeding depression has been documented. In 1999, Peter Crnokrak and Derek Roff reviewed every study they could find pertaining to the assessment of inbreeding depression in wild populations. That amounted to about 150 studies representing 35 vertebrate species—including Arabian oryx, harbor seals, rock wallabies, red-cockaded woodpeckers, desert top-minnows, great reed warblers, and more. Among those studies, more than half found evidence of inbreeding depression.

INBREEDING DEPRESSION IS VARIABLE AND DIFFICULT TO DETECT

If inbreeding and its adverse effects arise from such basic processes (Figs. 7.4 and 7.6), then why is inbreeding depression not detected more consistently than indicated by the work of Crnokrak and Roff? That question has two important answers.

First, inbreeding depression is stochastic in two essential ways. One can be appreciated by reviewing a detail of Figure 7.4. In that figure, the full-sib matings did not have to result in homozygotic offspring. It is possible (by chance) that the third generation of organisms in that pedigree could have all been heterozygotes.

A second element of stochasticity pertains to deleterious recessive alleles. To see how, suppose there is a large population in which inbreeding occurs at a slow rate. Further suppose that the population's habitat is fragmented, which results in several smaller subpopulations that are isolated from one another. Each subpopulation is now subject to a much higher rate of inbreeding. The severity of inbreeding depression in each subpopulation will depend on whether that subpopulation is harboring any deleterious recessive alleles. If there are no such alleles, then inbreeding depression may be mild to the point of being undetectable.[10] Furthermore, deleterious recessive alleles tend to be rare in a large population. So, when a large population is reduced and fragmented, it is possible that some subpopulations would end up (by chance) with fewer (or perhaps, no) deleterious recessive alleles. Other populations might end up (by chance) with more deleterious recessive alleles. The populations with more deleterious recessive alleles are said to have a larger *genetic load*. The severity of inbreeding depression depends on the size of a population's genetic load, which can be importantly influenced by stochastic processes. These ideas are powerfully supported by an experiment conducted on captive-raised populations of mice (Fig. 7.11).

Inbreeding tends to be detrimental. But its influence is often difficult to detect and varies considerably from case to case.

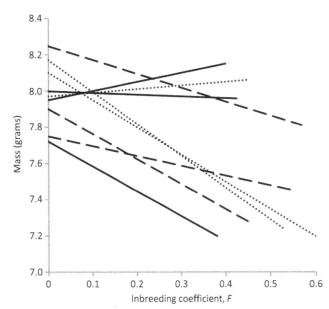

Figure 7.11 Severity of inbreeding depression for mean mass (grams) of offspring at the time of weaning for three replicate populations of three subspecies of beach mouse (*Peromyscus polionotus*). The three replicate populations belonging to each subspecies are indicated by the same line type (solid, dashed, and dotted). Differences among replicate populations occur, in part, because each lineage was founded by a few individuals and because—by chance—some lineages began with more deleterious recessive alleles (lethal equivalents) than other lineages. Source: Lacy, Alaks, and Walsh (1996).

Aside from inbreeding depression being a stochastic process, there is a second reason why Crnokrak and Roff did not detect inbreeding depression more frequently than they did. That is, inbreeding depression is difficult to detect, even when it is occurring. Detecting inbreeding depression requires measuring an individuals' F and some fitness-related phenotype. Those measurements are sometimes made too imperfectly (i.e., with too much measurement error), or sometimes too few observations are gathered. Either circumstance can lead to failing to detect inbreeding depression.

The wolves of Isle Royale illustrate this concern well. The population had long been known to experience high rates of inbreeding due to its small size, but it exhibited no detectable sign of inbreeding depression. For example, when analyses like those shown in Figure 7.9, panel A, and Figure 7.10 are applied to Isle Royale wolves, there is scant indication of inbreeding depression (Hoy et al. 2023). Because inbreeding depression had not been convincingly demonstrated, some well-respected scientists believed that the wolves of Isle Royale were a good example for how concerns about inbreeding are sometimes exaggerated.

For a long time, my colleagues and I had only faint clues, such as the population's decade-long delay in recovering from having collapsed after being exposed to a novel disease in the early 1980s (see also Fig. 9.5). Inbreeding depression was a plausible cause for the delay, but no more than plausible.

After 25 years of being on the lookout, we found the first significant evidence for inbreeding depression in 2009. We had been collecting the skeletal remains of Isle Royale wolves since the 1960s. But it wasn't until 2009 that we collected enough specimens and understood how to analyze them. We found that a kind of bone deformity, known as a lumbosacral transitional vertebra, had gone from quite rare to extremely common over the previous five decades as the population become more inbred (Räikkönen et al. 2009). This is the deformity exemplified by the Figure 7.2. Even this evidence was circumstantial, because we could show only that the incidence of the deformities increased over time; we were never able to more directly associate its incidence with the inbreeding coefficients of individual wolves.

INBREEDING DEPRESSION AND EXTINCTION

If inbreeding has an adverse effect on the individuals of a population, then it is likely to adversely impact the viability of the entire population (Vucetich and Waite 1999). While that idea is remarkably intuitive, it is also difficult to assess directly. Let's explore what's known about this idea.

Laboratory Populations

The basic idea is to raise inbred and outbred populations of *Drosophila* or bean beetle or some other small, easy-to-rear creature with a short generation time and look for differences in the time it takes for these populations to go extinct. While the idea is simple, experiments like this often involve several years and hundreds of vials or petri dishes containing portions of one of the experimental populations. To control the rate of inbreeding typically requires carefully moving individuals at just the right time from one vial to another to control their mating.

Tedious as these experiments are, quite a few have been conducted. A representative example can be seen in experiments with the work of *Drosophila* by researchers in Australia in the 1990s . They raised 10 to 30 replicate populations at each of 6 levels of inbreeding, including replicates that were outbred. Their experiment included one additional feature: they raised some populations in benign environments and other populations in stressful environments. The form of stress that they induced was either (depending on the replicate population) adversely high temperature (28.5°C instead of 25°C) or lacing the *Drosophila* food with ethanol, which represents the effect of eating fermented fruit and is detrimental to most *Drosophila*. They raised these populations for eight generations and observed which replicate populations went extinct. In the benign environments, the proportion of extinct populations was lowest for the outbred populations and greatest for the inbred populations (Fig. 7.12). In those benign environments, the

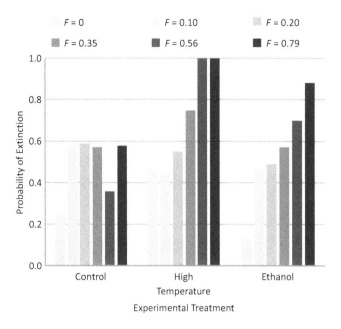

Figure 7.12 The probability of extinction after eight generations among replicate populations of fruit fly (*Drosophila*) raised under different levels of inbreeding and different environmental conditions. Source: Bijlsma, Bundgaard, and Boerema (2000).

rate of inbreeding didn't have much effect on extinction risk. In the stressful environments, rates of extinction were much higher, and there was a greater tendency for higher rates of inbreeding to lead to higher rates of extinction.

Wild Populations of Glanville Fritillary Butterflies

Aside from the wolves of Isle Royale, the most direct assessment of inbreeding and extinction in wild populations brings us to another island. Fasta Åland is a blob-shaped island about 40 km across, resting off the coast of southwest Finland. It's a mixture of rocky outcroppings, forests, and meadows. Many of the meadows are inhabited by two herbaceous species of plant—ribwort plantain (*Plantago lanceolata*) and spiked speedwell (*Veronica spicata*). They are attractive, if not a little weedy looking, with white and purple flowers (respectively). What's salient about these plants is that Glanville fritillary butterflies like to lay their eggs on these plants and only these plants. These diminutive butterflies have wings that span just an inch and a half and are decorated with a miniature checkerboard of beige and rusty orange patches. They flutter over meadows throughout much of Europe.

On Fasta Åland, a team of researchers found hundreds of meadows inhabited by ribwort plantain or spiked speedwell, each capable of supporting a small subpopulation of Glanville fritillaries. Because Glanville fritillaries tend not to travel far and because the meadows are far enough apart, many of these populations are more-or-less isolated from each other. Many of these subpopulations are also small, with the next generation rising from perhaps just 1 to 5 egg masses, where each egg mass consists of between 50 and a few hundred eggs that will hatch to become full siblings. Given those circumstances, it is common for a subpopulation to go extinct and about as common for an empty habitat patch to be recolonized by an adventurous dispersing butterfly.

Researchers identified 42 subpopulations of Glanville fritillaries on Fasta Åland, in a range of sizes and degrees of isolation from other subpopulations. They caught females from these sites in 1996 for the purpose of making laboratory-based estimates of heterozygosity. They also went back to the same sites a year later to see which subpopulations had survived. Seven went extinct. What they found is that populations with lower heterozygosity were more likely to have gone extinct (Fig. 7.13). That pattern was found after taking account of other potentially influential factors such as size of the habitat patch, its isolation from other patches, and size of population.

Projected Impacts of Inbreeding on Extinction in Wild Populations

By now it must seem obvious that *direct* assessment of inbreeding's impact on extinction risk for wild populations of longer-lived creatures is extremely difficult. An important approach to that difficulty is to assemble available information to develop a useful projection of the likely impacts. More specifically, the idea is to perform population viability analyses (PVAs) in a

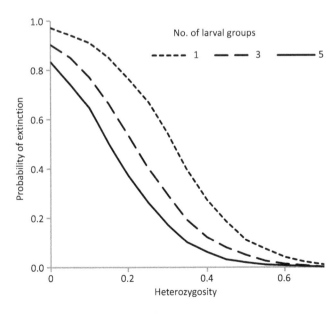

Figure 7.13 The relationship between proportion of heterozygotic loci and probability of extinction for subpopulations of Glanville fritillary butterflies. Each line is for populations supported by different numbers of larval groups. The lines resulted from fitting statistical models to data representing populations of different sizes and different levels of heterozygosity that did and did not go extinct. Source: Saccheri et al. (1998).

way that takes account of inbreeding and its impact. A software package such as VORTEX (discussed in the previous chapter) is sufficiently powerful and flexible to allow such an accounting.

A prerequisite for performing such a PVA is an ability to precisely quantify how vital rates, such as survival and reproduction, are influenced by different levels of inbreeding. To this end, recall panel A of Figure 7.9, where the slope of the line in that graph is an indicator for the severity of inbreeding depression. The y axis of panel A is mortality. It turns out (for reasons not explained here) that it is better to calculate this slope after converting mortality rate into a log-transformed survival rate, as represented by this equation for a straight line:

$$\ln[S] = a - bF, \tag{7.4}$$

where S and F represent the survival rate and inbreeding coefficient of an individual. In this equation, a is the intercept, which happens to be the logarithm of survival in the absence of inbreeding (i.e., $F = 0$). b is the absolute value of the slope of the relationship and an indication of the severity of inbreeding depression. The steeper the slope, the more severe the inbreeding depression. Equation 7.4 can be rewritten to focus on more specific vital rates, such as adult survival and juvenile survival, or different vital rates such as fecundity.

It turns out (again, by reasoning not shown here) that b is also an estimate of the expected number of lethal equivalents in an individual's genome—that is, lethal equivalents that might impact the trait of interest, such as adult or juvenile survival. In addition to having that informative interpretation, Equation 7.4 is key for incorporating the influence of inbreeding depression into a PVA. In a PVA that ignores inbreeding depression, the probability of survival for an individual might simply be whatever fixed number is substituted for S. But in a PVA that takes account of inbreeding depression, the probability of survival for an individual depends on their inbreeding coefficient, as specified by Equation 7.4. Taking account of inbreeding depression in a PVA also requires keeping track of who mates with whom to produce new offspring. Doing so allows a PVA to determine each individual's F. Conveniently, VORTEX is able to simulate and keep track of all those events and processes.[11]

Before using VORTEX in that way, it is necessary to have estimates for b (the number of lethal equivalents). A team of Australian researchers reviewed the published research on inbreeding depression for vital rates of wild populations (O'Grady et al. 2006). They found studies of 11 populations—representing, for example, tamarins, moorhens, elk, mice, mosquito fish, and several species of songbird—in which inbreeding depression had been studied for one or more of these vital rates: first-year survival, survival from one year of age to sexual maturity, and fecundity. Among those populations the average values of b were 2.4 for survival to year one, 6.0 for survival to sexual maturity, and 3.9 for fecundity.

Inbreeding depression tends to be more severe in wild populations and in more stressful environments (see also Figures 7.10, panel D, and 7.12).

The total number of lethal equivalents across an individual's life history is ~12. For context, the median number of lethal equivalents for cases of inbreeding depression in zoo populations is ~3.1 (Ralls, Ballou, and Templeton 1988). Those numbers can be interpreted as meaning that inbreeding depression is about four times more severe in the wild than in captivity.

The same team of researchers identified populations that had been studied well enough to perform a traditional VORTEX-style PVA—that is, studied well enough to have useful estimates of, for example, age-specific vital rates, density dependence, and environmental stochasticity. Then they performed PVAs under two conditions—namely, with and without taking account of inbreeding depression characterized by severities indicated by the abovementioned estimates of b.

The researchers identified 30 such populations. Because the rate of inbreeding is as important as the severity of inbreeding depression and because the rate of inbreeding depends on population size, the researchers performed simulations with different starting conditions (i.e., $N_o = 50$, $N_o = 250$, and $N_o = 1,000$). Regardless of the initial population size, the inclusion of inbreeding depression led to, on average and regardless of initial population size, a 37% reduction in median time to extinction. Figure 7.14 depicts the results for two representative populations. Interestingly, there was a tendency for the adverse impact of inbreeding depression on extinction risk to be greater for populations with positive growth rates as opposed to those with negative growth rates.[12]

EFFECTIVE POPULATION SIZE

Equation 7.1 describes precisely how the rate of inbreeding increases with decreases in N. Such equations are key for knowing how large a population needs to be to avoid the adverse effects of losing genetic diversity too rapidly. Recall that such equations were derived by making assumptions

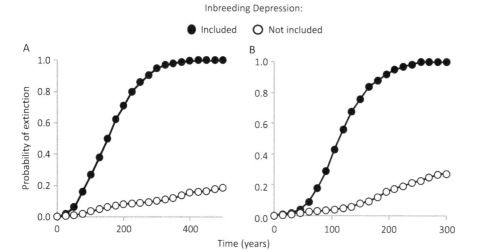

Figure 7.14 Probabilities of extinction for different times into the future for song sparrows (A) and African lions (B) for PVAs that did and did not include the effect of inbreeding depression. Source: O'Grady et al. (2006).

about how populations transmit alleles from one generation to the next—assumptions from which many real populations deviate. I'm referring to assumptions that rise from supposing that this transmission is like selecting marbles from a jar (Fig. 7.6). In particular, the jar-of-marbles analogy assumes that an offspring's two gametes were selected at random from the population. Compare that assumption with the behavior of a polygynous species, like southern elephant seals. In the Falkland Islands, one male may tend a harem of 50 or more females. Relatively few males in the population tend harems, yet these males end up siring about 90% of the offspring (Fabiani et al. 2004). As a result, gametes from the harem-tending males will be better represented in the next generation, as compared with those who do not tend harems.

While that example is extreme, it is common for some individuals to contribute more gametes to the next generation—more gametes, that is, than can be explained by the kind of chance represented by the jar-of-marbles analogy. If equations like 7.1 are to be useful, they need to be adjusted to take account of unequal contributions among parents to the next generation. These adjustments have been the subject of ongoing research for nearly a century, with the basic framework for making these adjustments having been developed in the 1930s.

To illustrate, recall that the heart and soul of Equation 7.1 is the realization that inbreeding increases at the rate $1/2N$, provided that the gametes for the next generation are selected at random from the parent population.

Join that idea to the expectation that a polygynous species such as elephant seals would tend to accumulate inbreeding more rapidly than $1/2N$ because so few males are contributing much in the way of gametes to the next population. To account for such a circumstance we need to rewrite Equation 7.1 by replacing N with N_e:

$$F_t = 1 - (1 - 1/2N_e)^t, \tag{7.5}$$

where N_e is called the *effective population size*. It has that name because it represents not the population's actual (census) size, but its effective size, given the realities of how it passes alleles to the next generation. The phrase "effective population" is also meant to contrast with an "ideal population," where an ideal population has that property of transmitting its alleles by randomly drawing from the alleles of the parent generation (as in Figure 7.6). Given those considerations, it is sensible, albeit very abstract, to define N_e as the size of an ideal population that experiences the same rate of inbreeding (or drift) as the population being studied (i.e., whose census size is N).[13] For context, N_e tends to be much smaller than N.

If, for example, a population is ideal in every way except the sex ratio of reproducing adults is skewed, then

$$N_e = (4N_m N_f)/(N_m + N_f), \tag{7.6}$$

where N_m and N_f are the number of males and females in the population. Suppose there is a population whose abundance (census size) is 1,000, and the sex ratio of reproducing adults is 8 females for every 1 male. In other words, there are 875 reproducing females and 125 reproducing males. Plug those values into Equation 7.6 and the result is 438. Then 438 can be plugged into Equation 7.5.

The most important ways by which a real population can deviate from an ideal population (in ways that tend to make $N_e < N$) are (1) uneven sex ratio of reproducing adults, (2) overlapping generations, (3) fluctuations in abundance over time, and (4) any behavioral or ecological process that leads to more variance in reproductive success among individuals than would be predicted by the random sampling process represented by the jar-of-marbles analogy. Virtually all populations have one or more of those four properties. A concrete example of that last property are situations in which some individuals have much higher quality home ranges, which in turn leads to greater reproductive success for those individuals.

Relatively simple equations—able to account for each of those four properties (one property at a time, like Equation 7.6)—have been available for decades. The ability to account for multiple properties as they occur in real populations, however, is extremely complicated business. On that account much has been learned gradually over the past 90 years. For example, in the 1990s, Leonard Nunney developed an equation that connects basic life-history information to an equation for N_e (Nunney 1996). The equation he developed is

$$N_e = (2NT) \, [2 + 2I_A + (2I_s/A) + I_{Zm} + I_{Zf} + I_A I_{Zm} + I_A I_{Zf}]^{-1} A^{-1}, \quad (7.7)$$

where N is the population's census size, T is the generation time, A is the average adult life span, and I is the standardized variance in fecundity (= [variance in fecundity among individuals]/[average fecundity per individual]^2) due to variation in adult life span (I_A), individual fitness (I_{Zm} for males or I_{Zf} for females), and random annual variation (I_s).[14] This equation accounts for many properties of real populations, except fluctuations in abundance. There is no need to follow the details of that equation. It's more than ample to appreciate that it is complicated and the culmination of decades of collective thinking on the topic.

I once used Equation 7.7 to estimate N_e for the wolves of Isle Royale. The resulting value was $N_e = 3.8$ at a time when the population's average size (over a decade or so) was approximately 24 wolves. Even though the wolf population had 24 individuals, it was losing genetic diversity as fast as an ideal population whose size would be a little less than 4. Very soon, we will find it interesting to have noted that the ratio of N_e/N for this example is ~0.16 (= 3.8/24).

THE N_e/N RATIO

Because the data required to estimate N_e is difficult to obtain, it is not possible to estimate N_e for most populations of conservation concern. For that reason, there has been considerable effort to search for general patterns among well-studied populations that might apply to less-studied populations.

The main strategy in this search involves recognizing that N_e is typically much less than N. Perhaps there is a common range of ratios (N_e/N) for wild populations. For example, and only hypothetically, if N_e/N were typically ½, then one could estimate N_e as half the population's census size.

After studying N_e/N ratios for scores and scores of populations, here's a summary of what's known: N_e/N varies widely among populations. Nevertheless, N_e/N is rarely greater than 0.5 and commonly between 0.10 and 0.25, meaning that N_e is commonly between a quarter to one-tenth of the population's census size. That previous statement is not—by itself— especially useful as universally applicable guidance, because much variation in N_e/N is attributable to taxonomic differences. Specifically, creatures characterized by high fecundity, such as marine fish, tend to have lower values of N_e/N (often lower than 0.01), and creatures characterized by low fecundity tend to have higher value of N_e/N, sometimes higher than 0.30. Consequently, if there is an interest to estimate a plausible range of values for N_e/N for some specific population of conservation concern, it would be best to review what's known about N_e/N for the taxon to which the species of concern belongs.

> While the ratio N_e/N varies widely among populations, it is commonly between 0.10 and 0.25. Furthermore, much variation in N_e/N is attributable to taxonomic differences, with N_e/N being much lower for species with high fecundity.

There is also evidence suggesting that N_e/N may tend to be lower for populations with smaller N (Luikart et al. 2010). Finally, some variation in N_e/N may also be due to differences in methodology and degrees to which populations deviate from the assumptions on which the employed estimator depends.

MINIMUM VIABLE POPULATION SIZE, REVISITED

We thought about minimum viable population size (MVP) in Chapter 6 without taking account of genetic processes. Now it's time to ask, How large does a population need to be to maintain its genetic health? In so asking, we'll focus on two aspects of genetic health in each of the next two sections.

Inbreeding and Genetic Health

We know much about inbreeding from math-based theory and formal empirical research, by which I'm referring to the kinds of knowledge discussed throughout this chapter. But there is also an important body of knowledge coming from the cumulative, direct experience of animal breeders, including people who breed animals in zoos, for agriculture, and recreational pursuits (such as horse racing and owners of pets, such as dogs, pigeons, and fish). From well over a century of such experience, breeders have long had the sense that animals have health issues when inbreeding coefficients increase by ~2% or more per generation. From that experience, a 1% rate of increase has been judged as guidance for a safe maximum rate of inbreeding.

To find the value of N_e that corresponds to that presumed safe limit, take the equation $\Delta F = 1/2N_e$, replace ΔF with 0.01, and solve for N_e. The result is $N_e = 50$. Suppose that you have reason to think N_e/N for the population of concern is 0.25, then take the safe limit of N_e (50) and divide it by 0.25. The result is $N = 200$.

Frustratingly, laboratory experiments with invertebrate species suggest that $N_e = 50$ is not large enough to avoid inbreeding depression (Latter et al. 1995; Bryant et al. 1999; Reed and Bryant 2000). The best-available guidance for how large wild populations need to be to avoid inbreeding depression may well come from analyses of captive invertebrates raised in stressful environments—like those represented by Figure 7.12.

Maintaining Adaptive Genetic Diversity

Inbreeding is only one concern pertaining to genetic health. Another concern is the maintenance of adaptive genetic diversity. To appreciate the nature of this concern, recall the bathtub metaphor earlier in this chapter, where we emphasized that genetic diversity is lost at some rate by some processes (drift, inbreeding), added at some rate by other processes (mutation), and maintained by other processes (selection).

The begged question is, How large does N_e have to be so that genetic drift is a sufficiently weak force? Weak enough, that is, for natural selection to maintain beneficial alleles and eliminate mutations that end up being deleterious. The mathematical theory required to derive answers to that question is not beyond your ability, but it is beyond the scope of this book. Nevertheless, we are interested to note the kind of answers that have emerged from such considerations. In a nutshell, some have provided reasoning that suggests N_e should be no less than 500 (Franklin 1980), while others have provided evidence indicating that N_e should be no less than 1,000 to 5,000 (Lande 1995; Lynch and Lande 1998). Those safe limits correspond to census sizes on the order of 3,000–30,000 animals in a population, assuming that N_e/N is typically on the order of 0.15.

POPULATION FRAGMENTATION AND GENE FLOW

We have been focusing on the population genetics of single, isolated populations. But many populations exist as a set of subpopulations that are largely (but not entirely) isolated from one another. The genetics of these populations differ. You will quickly intuit this difference by recalling Figure 7.7. We first imagined each line of those graphs as representing different hypothetical outcomes for a single population, as if we watched the population's dynamics play out, then rolled back time, and watched it unfold again.

Now, imagine panel A of Figure 7.7 to represent a large population (200 individuals) fragmented into 10 smaller subpopulations, each comprised of 20 individuals. In other words, let each line in panel A represent 1 of the 10 populations. Genetic drift causes all the subpopulations to lose genetic diversity, but some populations lose one allele, and other populations lose the other allele. If immigrants were to occasionally move between the subpopulations, they would likely bring with them a copy of an allele type that is rare in the subpopulation to which they arrive. If immigration occurred at just the right rate, then the loss of diversity within any subpopulation might be prevented or delayed. This movement of alleles is referred to as *gene flow*.

You won't be surprised to know that an equation was developed (again, back in the 1930s) to predict how different levels of gene flow affect a subpopulation's F. The equation is

$$F_{ST} \approx 1/(4N_e m + 1), \qquad (7.8)$$

where *m* is the migration rate, or the proportion of the recipient subpopulation that is comprised of immigrants at each generation.

You will no doubt notice that F appears in Equation 7.8 with the subscript "ST" (as it did in Figure 7.3). Now, it's time to share a fuller truth. F is not a single idea, but a constellation of ideas and a set of statistics. There are three F-statistics whose purpose is to describe the distribution of genetic diversity within and among populations (of the same species):

- F_{IS} is the probability that two alleles—from one locus in one individual—are identical by descent. This is the idea first described in the first paragraph of the section entitled "Inbreeding Coefficients."
- F_{ST} describes the effect that population fragmentation has on genetic drift. It is the probability that two alleles, drawn at random from a population fragment, are identical by descent. That probability increases as a population fragment becomes increasingly isolated from other fragments.
- F_{IT} is the total inbreeding and takes account of both F_{IS} and F_{ST}. We won't concern ourselves any further with F_{IT} in this book.[15]

In a small, isolated population, F_{IS} and F_{ST} both tend to increase over time. And that rate of increase for both statistics is usefully described by Equation 7.5.

Now, let's refocus on Equation 7.8. Notice that the equation does not include any measure of time. The reason is that gene flow causes F to equilibrate at some value less than one, and Equation 7.8 tells us about that equilibrium value. Furthermore, geneticists have concluded that an F_{ST} of ~0.2 represents a healthy level of diversity among population fragments (Mills and Allendorf 1996). If F_{ST} is too much higher, then loss of genetic diversity is likely to be a problem. If F_{ST} is too much lower, then population fragments risk becoming too genetically homogenous—perhaps to the point of not being able to maintain local adaptations. The appropriateness of values tending toward $F_{ST} \approx 0.2$ is not proven by any formula; rather, it is a judgment of well-qualified geneticists.

Given this target ($F_{ST} \approx 0.2$), there is a simple, albeit clever, math maneuver to perform. Specifically, replace F_{ST} on the left side of Equation 7.8 with 0.2, and solve that equation for the product, $N_e m$. You can easily confirm that the result is $N_e m \approx 1$. Furthermore, whereas m is a *rate* of migration, the product $N_e m$ is the *number* of (effective) migrants each generation required to result in $F_{ST} \approx 0.2$. The actual number of migrants is found by dividing $N_e m$ by the N_e/N ratio. So, if the N_e/N ratio is 0.25, then the actual number of migrants per generation is approximately 4.

The result is convenient because it means that F_{ST} depends on the number of migrants per generation, no matter the population's size. The result is also remarkable for indicating that a good balance between maintaining diversity and avoiding homogenization is realized by a small amount of gene flow. This result has been dubbed the *one-migrant-per-generation principle*, even though the actual number of migrants required is likely to be anywhere from 1 to 10, due in large part to N_e/N being so much less than 1.

While appropriate levels of gene flow among population fragments is a critical element of conservation, very few recovery plans consider it. A rare example of gene flow being taken into account are wolves of the Northern Rockies of the United States. These wolves are divided into three subpopulations (central Idaho, northwest Wyoming, and northwest Montana). Official recovery planning requires considering this population endangered if there is inadequate gene flow among the subpopulations (Hebblewhite, Musiani, and Mills 2010).

Another example of considering the need for gene flow is the work of Kenney et al. (2014), who used VORTEX to assess the need for migration between population fragments of tigers. For some population fragments, more than four immigrant tigers per generation were required to stave off inbreeding depression.

GENETIC RESCUE

The aforementioned ideas are valuable for maintaining genetic health and preventing inbreeding. However, conservationists increasingly find themselves facing situations in which inbreeding depression is an immediate threat and in need of being mitigated. In such cases, genetic rescue has become an important prospect. The idea here is that the adverse effects of inbreeding on population growth rate and extinction risk can be reversed by introducing just one or a few individuals into a population.

Genetic rescue has been demonstrated with experiments conducted in labs on a number of occasions (Frankham 2015). Genetic rescue has also been observed in about a dozen cases for wild populations (Whiteley et al. 2015), including bighorn sheep, adders, prairie chickens, mountain pygmy possum, wolves, and Florida panthers. Most of those situations were similar: the population was not doing well, inbreeding depression was documented or suspected, an immigrant arrived either on its own or with the help of humans, and the population subsequently improved (Fig. 7.15).

As impressive as such cases have been, genetic rescue is still surrounded by several questions, such as what exactly are the genetic mechanisms by which genetic rescue works, how long-lasting (or short-lived)

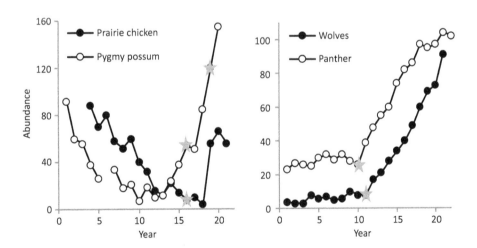

Figure 7.15 Examples of genetic rescue. These examples show how the introduction of one or a few individuals seems to have resulted in significant increases in abundance due to an improvement in the population's genetic health. Stars mark the times when immigrants arrived to the population. Sources: Madsen et al. (2020), Westemeier et al. (1998), Johnson et al. (2010), and Vila et al. (2003).

is its benefit, and can the introduction of an immigrant ever be detrimental (Bell et al. 2019)?

Genetic Rescue and the Wolves of Isle Royale

Each year, field work for studying the wolves of Isle Royale involves many miles of off-trail hiking and visiting dozens of sites where wolves had killed moose. In doing so, we come across scads of wolf scat. We began collecting samples of these scats in earnest back in the late 1990s—stock-piling them in freezers on the prospect that one day we could afford to analyze them for their DNA. By 2011, we'd done so and used the results to build a pedigree. The pedigree revealed—much to our surprise—that a wolf had immigrated to Isle Royale from Canada by crossing an ice bridge that formed in 1997 and connected Isle Royale to the mainland. While we had not previously known this wolf's heritage, we had long known his identity—he was larger and more aggressive than most wolves. He was one of the most successful wolves (in terms of lifetime reproductive success) to ever live on Isle Royale. He also turned gray as he aged. That graying was a phenotype not previously observed in an Isle Royale wolf. We grew into the habit of calling him the Old Gray Guy.

The pedigree also allowed us to trace the ancestry of this immigrant wolf. Ancestry is a measure of the portion of genes in a population that trace to an individual. Within 2.5 generations of the immigrant's arrival,[16] his ancestry grew to ~60%. In other words, 60% of the population's genome was inherited from this one wolf. That meteoric rise in ancestry is direct evidence that the immigrant and his descendants were that much more fit than the lineage of native wolves. That finding is also the first *definitive* evidence that the population had exhibited inbreeding depression.

As powerful as the genetic rescue had been, there was no strong increase in abundance, like that observed in other cases (compare Figs. 7.15 and 9.5). The likely reason is that moose abundance had collapsed (due to a catastrophically severe winter) a year prior to the immigrant's arrival. That collapse in moose abundance meant that foraging was more difficult for wolves than usual. Had the immigrant not bolstered the population in the way that he did, the food shortage likely would have pushed the population to extinction in the late 1990s.

The benefits of this genetic rescue were also short-lived. Within a few generations, much inbreeding had occurred within the immigrant's lineage (Fig. 7.5), and the population was on the brink of extinction. Park managers waited until the population dwindled to the last 2 wolves (middle and left wolves in Figure 7.1), then translocated about 19 wolves to Isle Royale. Some of the translocated wolves killed that last male, and the last female died without reproducing again. In that sense, the population went extinct and an entirely new population was started.

Finally, we had wondered, if the benefits of an immigrant were so important, yet short-lived, how had the population persisted since it was founded (in about the year 1950)? You can discover the answer to that question through an exercise included with the online supplement.

CHAPTER 8

Exploited Populations

You did not kill the fish only to keep alive and to sell for food, he thought. You killed him for pride and because you are a fisherman. You loved him when he was alive and you loved him after. If you love him, it is not a sin to kill him. Or is it more?

—Ernest Hemingway, *The Old Man and the Sea*

The exploitation of wildlife is multifaceted. Modes of exploitation include hunting, trapping, fishing, and removal of live animals for the pet trade. Exploitation is also purpose-driven and motivated by a variety of reasons, such as recreation, subsistence, and commerce. Exploitation is often of vital importance to human well-being and often not. Exploitation may be legal or illegal. Illegal exploitation should not be assumed as having no purpose or reason. Typically, it does. Legal exploitation may be highly regulated, loosely regulated, or essentially unregulated. Legal exploitation is sometimes well-regulated and other times not. Well-regulated exploitation should not be confused with highly regulated exploitation, because a harvest with little regulation can sometimes meet its goals, and if a harvest meets is goals, then it is well-regulated. And a regulated harvest can be poorly regulated, if the regulations are set poorly. Many populations of wildlife are exploited without that exploitation being a threat to conservation; at the same time, exploitation is too often a grave threat to conservation. It bears repeating that the exploitation of wildlife is multifaceted.

The exploitation of wildlife is also extensive. To illustrate, consider the Convention on International Trade in Endangered Species (CITES), which is an international treaty designed to protect species from going extinct due to international trade in the products that result from exploiting populations. CITES's aim is not prohibition of exploitation, but rather the promotion of sustainable exploitation. An indication of the extent of wildlife exploitation is that CITES calls for some level of protection to nearly 6,000 animal species.

BENEFITS TO HUMAN WELL-BEING

One of the many facets of exploiting wildlife is the benefit it brings to human well-being, which is immense and wide-ranging. For example, indigenous communities living on the shores of the Arctic Ocean and Bering Strait depend culturally and nutritionally on subsistence hunting of marine mammals. For some of these communities, marine mammals represent 80 kg of food per person per year (Alaska Department of Fish and Game 2014).[1]

Bushmeat is a critical source of protein for an estimated 150 million rural households in Africa, south Asia, and South America. In rural regions of the Congo, humans are at greater risk of malnutrition when overharvesting leads to depleted populations of wildlife (Fa et al. 2015). The same circumstance leads to higher rates of anemia in children living in some rural areas of Madagascar (Golden et al. 2011).

About 10% of all humans' livelihoods are supported by fishing, and fish represent 20% of daily protein intake for some 3 billion humans (Sustainable Fisheries 2022).

Aside from commercial and subsistence harvest, much exploitation is conducted for recreational purposes. Finally, the purpose of some exploitation is to reduce a population's abundance to realize some perceived benefit of doing so. An example would be to reduce predator abundance as a means of increasing hunters' satisfaction.

THREATS TO CONSERVATION

Overexploitation has led to more than 4,800 vertebrate species being classified as threatened or near threatened according to the International Union for the Conservation of Nature (IUCN) Red List Criteria. That number is stunning. Among bird and mammals, which are well-studied taxa, at least ~16% of those species are threatened or nearly threatened by overexploitation (Maxwell et al. 2016).

The threat of overexploitation varies geographically. For example, overexploitation is the number one cause of endangerment in China, with

78% of that nation's endangered species being so because of overexploitation (Yiming and Wilcove 2005). By contrast, overexploitation is the fourth most important cause of endangerment in the United States, behind habitat loss, pollution, and non-native species. While lower in the list of causes, overexploitation is a cause of endangerment for 27% of that nation's endangered species. The high frequency of overexploitation in China is tied to the larger number of poor rural people who live near wildlife and can readily exploit wildlife as a way of sustaining themselves.

While exploitation is a common threat to conservation, many populations are exploited without any direct threat to conservation.

PURPOSE AND CONFLICTING INTERESTS

The aforementioned ideas are a tiny sample from a cornucopia of benefits and threats that arise from exploiting wildlife. What can unify this diversity of circumstances is knowing that exploitation is a purpose-driven activity, in which the purpose should be to deliver some specified and broadly agreed upon benefit without incurring too many of the detriments. The concern, however, is that purposes of exploitation vary widely. Thus, we will do well to provisionally consider a stylized purpose—namely, to harvest a *large and consistent* number of individuals from a population each year without reducing population abundance by too much or causing an increase in the risk of population collapse.

BEST PRACTICES AND COMMON PRACTICES

The best way to realize any set purpose of exploiting wildlife includes analysis like that outlined below. Before you even read about this best practice, I can tell you that actual practices routinely fall short. They fall short because of a felt need to harvest, while lacking one or more of the following: adequate data about the exploited population, expertise among managers, control of hunters' behavior, or sometimes sufficient interest. The risks incurred by giving short shrift to best practices are easiest to appreciate by first understanding the population biology of exploited wildlife populations. Let's go.

UNSTRUCTURED POPULATIONS

In the following sections, we'll consider how unstructured, density-dependent populations (Chapter 3) are affected by different *harvest strate-*

Figure 8.1 Willow ptarmigan. Source: https://upload.wikimedia.org/wikipedia
/commons/1/1c/Lagopus_muta_%28Bernese_Oberland%29.jpg.

gies. A harvest strategy refers to how the intensity of harvest varies with population abundance.

A Case Study

The population biology of exploitation can be easier to apprehend if one has a real case in mind. Let's take willow ptarmigan as a case (Fig. 8.1). They live in thin forests and thickets across the world's subarctic and subalpine habitats. They have a very high r_{max} (~1.5/year), in part due to females being able to produce each year, on average, 3.7 fledglings that survive to the fall hunting season. But ptarmigan reproduction also varies considerably from year to year, resulting in marked fluctuations in abundance. Important contributors to these annual fluctuations include weather and fox predation.

Fluctuations in reproduction contribute to large fluctuations in abundance. Population highs are commonly 2.5 to 4 times higher than population lows. Across the landscape, ptarmigan density can range from 5 to 25 birds per square kilometer. Because willow ptarmigans are difficult to count accurately, there is often limited knowledge about whether a local population is more or less abundant than typical at any point in time.

In recent decades and at many sites in the United Kingdom and Fennoscandia, willow ptarmigan populations have declined dramatically. For example, the number of ptarmigans taken annually by hunters in two northern counties of Norway fell from 107,000 to 38,000 between 2000 and 2016 (Breisjøberget et al. 2018a).[2]

Willow ptarmigan have been exploited for centuries. In many places, ptarmigan hunting has also long been largely unregulated, in the sense that there is typically no legal limit to the number of hunters or bag limit per hunter (daily or annual). Furthermore, in Fennoscandia landowners can decide how intensely to hunt on their lands. While causes of ptarmigan decline are unknown, candidate explanations include habitat degradation due to over-browsing by moose and disrupted rodent cycles (due possibly to climate change), which lead to increased predation on ptarmigan by foxes (Henden et al. 2011; Breisjøberget et al. 2018b). Hunting is thought not to be a cause of decline by most who study the problem. But the declines have been dramatic enough to cause some of these same experts to think that more restrictive hunting is warranted. So, what harvest strategy and what intensity of harvest are most appropriate?

Constant Quota Strategy (CQS)

Suppose that the ptarmigan population on some landowner's property is well described by linear density dependence with $K = 1,000$ and $r_{max} = 0.2$/year.* Further suppose that 45 individuals are harvested each year. If so, then next year's abundance would be given by

$$N_{t+1} = N_t + N_t r_{max}(1 - N_t/K) - H. \qquad (8.1)$$

This equation is identical to Equation 3.3, except it also includes a term ($-H$) to describe the number of individuals lost to harvest each year. H is sometimes referred to as a quota, and the idea behind Equation 8.1 is that the quota is constant, no matter what the population's abundance might be. One challenge with such a strategy is to find a value of H that is large

---

* This value of r_{max} is unrealistically low for ptarmigan, but useful for pedagogical purposes.

enough to satisfy the demand for harvest, but not so large as to risk over-exploitation. If an appropriate value of H can be determined, this strategy might seem to have two positive attributes: the harvest return is constant from year to year and the harvest can be implemented without needing to know N_t each year, because H is the same every year, no matter the value of N_t.

> There is a strong benefit in building this spreadsheet now, before reading on.

To gain the best understanding of exploitation's impact on populations, you'll want to make your own calculations in Excel. Specifically, use Excel and Equation 8.1 to project the abundance of 10 populations over a 50-year period. Let each of these 10 populations have a different initial abundance (100, 200, . . . 1000). These projections lead to three observations (Fig. 8.2):

· None of the populations equilibrates at K.
· The seven populations that started with higher abundances all equilibrate to ~660 individuals.
· The three populations that started with lower abundances went extinct.

In Excel, reduce the annual harvest from 45 to 20 individuals per year. What happened? Fewer populations went extinct, and the ones that survived equilibrated to a higher value.

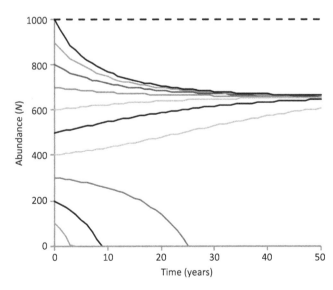

Figure 8.2 Projected abundances of 10 harvested populations that are identical except for the starting abundance. The populations' carrying capacity is marked with a dashed line at $N = 1,000$.

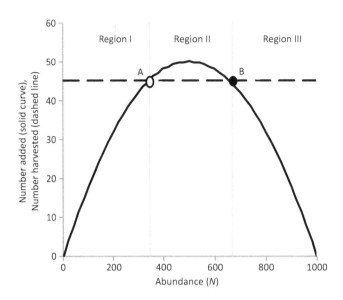

Figure 8.3 Dynamics of an exploited population where a constant number of individuals are removed. The solid line represents annual population growth due to density-dependent processes, and the dashed line represents annual losses due to harvest. The meanings of the labeled points (A, B) and regions of the x axis (I, II, III) are provided in the main text. The dashed line may be thought of as a harvest strategy, which in this case, is to harvest 45 individuals no matter what the population's abundance.

The results in Figure 8.2 motivate the development of a flexible framework for better understanding exploitation's impact on populations. This framework includes a graph that depicts two lines (Fig. 8.3). One line shows the harvest strategy—that is, the *number of individuals lost each year* due to harvest as a function of abundance (N). In this example, this is a horizontal line whose y intercept is 45.

The other line on this graph shows the *number of individuals added each year* as a function of N, due to density-dependent processes. To make this line, slightly rearrange Equation (8.1) by subtracting N_t from both sides of the equation, yielding

$$N_{t+1} - N_t = N_t r_{max}(1 - N_t/K) - H. \quad\quad (8.2)$$

Think of the left side of the equation as a single quantity, interpreted as the population's change in abundance over the upcoming year or *growth rate*. The left side of Equation 8.1 is distinct from the *per capita growth rate* (which is r_t).

To recreate Figure 8.3, create a column of numbers in Excel representing abundance (0, 20, ... 1000), which will be the graph's x axis. Create a second column representing the growth rate $(N_{t+1} - N_t)$ that would occur if there were no exploitation. In other words, values in this second column are equal to $N_t r_{max}(1 - N_t/K)$, where r_{max} and K are fixed parameters and the values of N_t are those appearing in the first column of your spreadsheet. Finally, create a third column representing annual losses due to exploitation, which is 45 individuals for the case that we've been considering. The second and third columns are values for the graph's y axis. The graph of these columns should look like Figure 8.3.

Again, now is a good time to make this graph. Then see how the graphs depicted in Figures 8.2 and 8.3 are affected by different values of H, say 20, 50 and 55.

Figure 8.3 makes sense of our observations for Figure 8.2. To see how, recognize that where the two lines of Figure 8.3 cross, the additions and losses balance and abundance does not change. Specifically, notice the labeled points (A and B), where the lines cross at $N \approx 340$ and $N \approx 660$, respectively. If the population is at either abundance, then abundance won't change. These are equilibrium values. Let those two values of N divide the x axis into three regions (I, II, and III):

- If N_t is in region III, then the dashed line (losses due to exploitation) is greater than the solid line (gains due to density-dependent processes), and N_t will decrease year after year until N_t reaches ~660.
- If N_t is in region II, then gains exceed losses, and N_t grows each year until N_t reaches ~660.
- If N_t is in region I, then losses exceed gains, and N_t declines, year after year, until the population goes extinct.

The value at point B, $N \approx 660$, is a *stable* equilibrium, because if a population is at that value, it will remain there. If a stochastic event nudges abundance up or down, it will return to that equilibrium value. But point A, $N \approx 340$, is called an *unstable* equilibrium because a population at that value will remain there, but only if not perturbed by a stochastic event. Any slight perturbation will send abundance away from that equilibrium point. An upward perturbation will send the population to $N \approx 660$, and a downward perturbation will send abundance to extinction. That is, $N \approx 340$ is a threshold below which the population will go extinct.

Before analyzing this harvest strategy any further, we'll benefit from considering a different harvest strategy to form a basis for comparison.

Constant Proportion Strategy (CPS)

This strategy involves harvesting some percentage of a population each year (say 10%), no matter the population's abundance. By keeping the percentage constant, fewer individuals are harvested when abundance is low. This strategy is represented by Equation 8.1, except H is replaced with $N_t h$, where h is the proportion of the population harvested each year and $N_t h$ is the number harvested in any particular year. A graphical representation of this strategy is given in panel A of Figure 8.4, where h happens to be set to 0.069/year (i.e., about 7% per year). Increasing h has the effect of making the dashed line in panel A steeper.

Comparing Strategies

There are important points of comparison between CQS and CPS. First, the CPS example that we're considering, $h = 0.069$/year, leads to a stable equilibrium value of $N \approx 660$. For that abundance, the annual harvest is approximately 45 ($\approx N_t h = 660 \times 0.069$). That same outcome is predicted for the CQS when H is simply set to 45. In broader terms and for both strategies, the population's stable equilibrium is determined by H (for CQS) or h (for CPS). Higher values of either H or h lead to lower values of the stable equilibrium, and vice versa.* If either H or h is set too high, the population will be driven to extinction. In these regards, the two strategies are comparable.

Yet, an important difference between the strategies is that the number of harvested individuals fluctuates from year to year for the CPS. These fluctuations occur because in real populations N_t fluctuates from year to year, which causes H ($= N_t h$) to fluctuate. By contrast, the number of harvested individuals is constant from year to year for CQS, so long as the population persists.

Another important difference between these strategies is the risk of population collapse. The risk, more precisely, is that N drops below the lower (unstable) equilibrium of the CQS, after which the population is driven to extinction (Fig. 8.3). If H is set low, then this risk may be small. By contrast, the CPS does not have an unstable equilibrium (Fig 8.4, panel A). Consequently, the CPS allows for moderately high levels of harvest (by setting a moderately high h) without incurring too much risk of population collapse. The risk of collapse is reduced because the number of harvested

---

*To make sure you understand this statement, focus on Figure 8.4, panel A, and imagine reducing h by decreasing the slope of the dashed line. Then, note how the equilibrium value for N increases.

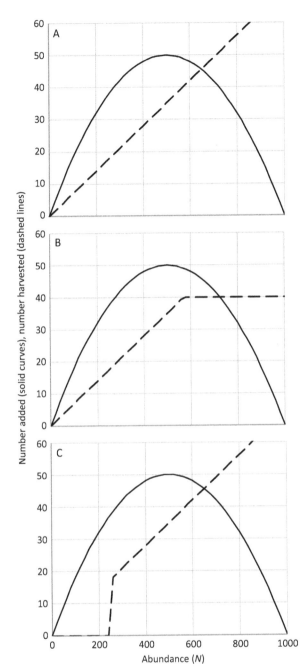

Figure 8.4 Loss functions (dashed lines) for three harvest strategies in relationship to gain functions (solid lines) for populations with linear density dependence ($K = 1000$, $r_{max} = 0.20$/year). The harvest strategies go by the names *constant proportion* (A), *restricted effort* (B), and *threshold* (C). See the main text for additional explanation.

individuals ($N_t h$) declines as N declines under the CPS. In this regard, CPS is considerably safer than CQS.

Reflecting on Those Comparisons

We've got more to learn about these strategies, but now is a good time to begin considering how they might apply to a specific case like willow ptarmigan. More specifically,

- CPS may be less risky, but risk is only one feature of a harvest strategy. Furthermore, the risk of CQS might be mitigated by setting a low H. For a case like willow ptarmigan, which strategy would likely be better? Why?
- If you had good population data (to make the solid curves in Figures 8.3 and 8.4), how would you go about setting H or h? Is that question answered, perhaps by asking, at what point does the detrimental effect of reducing ptarmigan abundance outweigh the value of increased harvest return? Any answer to that question is essentially an ethical claim.
- In terms of managing the behavior of hunters, would you guess that it's easier to control a constant value of H or h from year to year? Why?
- Would your answers to the preceding questions change if the case were not the recreational hunting of willow ptarmigan, but some kind of wildlife exploitation for commerce or subsistence? If so, how, and why?

Now is the best time to break from reading and record your thoughts on these questions. After that, you can discuss them with your classmates and instructor.

Implicit Assumptions

Our assessment of harvest strategies has, thus far, made four implicit assumptions:

- The availability of accurate estimates of N_t, r_{max}, and K to estimate the solid curves in Figures 8.3 and 8.4.[3] An estimate of N_t is needed to estimate each year's harvest under CPS and to monitor the impact of harvest under CQS.
- An appropriate population model is selected. For example, should one be using a structured population model like that described in Chapter 5, rather than the unstructured models we've considered so far?
- Hunters' effort and efficiency can be controlled so that safe limits for H or h are not exceeded. An example for how serious this concern can be is when the Wisconsin Department of Natural Resources (DNR) set a quota of 119 wolves in 2021. But 218 wolves were harvested. Many hunters may have used social media to

deceptively manipulate the reporting system that would have closed the season as soon as the quota was reached. The DNR was also criticized for using a reporting system that was easy to manipulate (Durkin 2021).

· The population is not influenced by environmental stochasticity.

An adequate understanding of harvest dynamics depends on tending to all of those ideas. Let's begin with environmental stochasticity.

ENVIRONMENTAL STOCHASTICITY

You learned much about environmental stochasticity in Chapters 3 and 6. Consequently, this section is focused on applying that knowledge to the development of a spreadsheet that assesses the impact of harvest strategies on stochastic populations.

We'll begin by building two graphs. One graph will show abundance over a 50-year time period for each of 2 populations, one subjected to the CQS and the other to the CPS. The second graph will show how the harvest return (number of individuals harvested) fluctuates from year to year for each population. The set up for this analysis is shown in Figure 8.5.

	A	B	C	D	E	F
1		Population parameters			Harvest parameters	
2	K	N_o	r_{max}	std dev of r_t	constant quota	constant effort
3	1000	900	0.2	0.1	45	0.069
4						
5						
6			CONSTANT QUOTA		CONSTANT PROPORTION	
7	year	N	harvest	N	harvest	
8	0	900		900		
9	1	925.5	45	770.9	62.1	
10	2	873.9	45	748.7	53.2	
11	3	813.1	45	802.2	51.7	
12	4	774.2	45	936.2	55.4	
13	5	845.5	45	771.7	64.6	
14	6	820.4	45	643.4	53.2	

Figure 8.5   A set-up for a spreadsheet that analyzes different harvest strategies in the presence of environmental stochasticity. Row 3 contains parameters describing the population and the harvests. These values were also used to make Figure 8.3 and panel A of Figure 8.4. The values in row 8 are initial conditions. The values in rows 9 through 14 are the result of formulas typed into those cells and described in the main text.

Now is a good time to make the spreadsheet depicted in Figure 8.5.

Writing the formulas for cells in row 9 requires some care and concentration. But once those cells are completed, they are simply copied and pasted into the cells below. The formula for cell <B9>, along with some annotations is

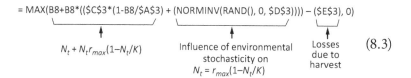

$$= MAX(B8+B8*((C3*(1-B8/A3) + (NORMINV(RAND(), 0, D3)))) - (E3), 0)$$

| $N_t + N_t r_{max}(1-N_t/K)$ | Influence of environmental stochasticity on $N_t = r_{max}(1-N_t/K)$ | Losses due to harvest | (8.3) |

Equation 8.3 makes use of the $= MAX(x, y)$ function in Excel, which tells Excel to report the larger of the two values, x of y.[4] In this particular case, it tells Excel to report the predicted abundance, unless that predicted abundance is negative, in which case it reports $N = 0$. The formula for <C9> is $= IF(B8 > E3, E3, B8)$. This formula reports the annual harvest return, which is either the set quota (H), or N_t if the harvest drives the population to such small values that N_t is less than the quota, by which time the population will soon be extinct and the harvest will be zero.

> What you read here will have limited meaning and learning value unless you implement these ideas yourself in Excel. Doing so will also make subsequent reading more meaningful.

The formulas for cells <D9:E9> are similar to cells <B9:C9>, except that the expression E3 in Equation 8.3, which represents losses due to harvest, is replaced with D8*F3, which represents $N_t h$. When you create this spreadsheet, observe the time delay. Specifically, the number to be harvested each year is based on abundance during the previous year. A reason to acknowledge such delay is that for many situations the best basis for estimating abundance at the time of harvest would come from the previous year. Careful analyses of harvesting white-tailed deer and moose show that such time delays likely make populations more variable than they would otherwise be (Fryxell et al. 2010).[5]

Because Equation 8.3 includes random values, each stroke of the F9 key on your computer produces a different realization of the dynamics. Get a feel for how these populations behave with different levels of harvest (H or h) and different levels of environmental stochasticity (which is set by cell <D3>).

Comparing Strategies for Stochastic Environments

With the aforementioned ideas, one can design a spreadsheet assessing many replicate populations exposed to a range of harvest intensities (H or h). An example of the results to come from such an assessment is given in Figure 8.7.

One way to interpret Figure 8.7 is to suppose one wanted the largest average annual harvest possible without there being any risk of extinction.

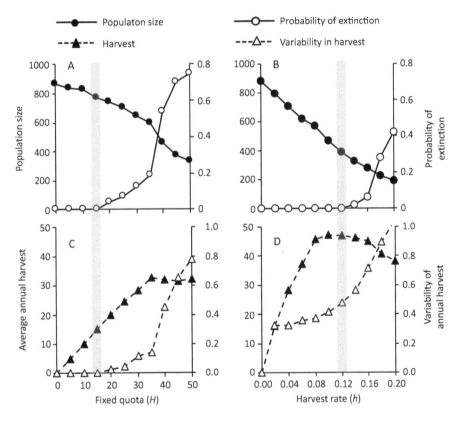

Figure 8.7 Dynamics for populations exploited according to different harvest strategies. The strategies depicted here are *constant quota* for a range of quotas (H) and *constant proportion* for a range of harvest rates (h). These strategies are applied to a population whose demography is characterized by $K = 1,000$, $r_{max} = 0.2$, and Var[r] = 0.04. The vertical gray bars highlight the highest levels of H and h that do not involve any risk of extinction within 50 years. The left axes of panels A and B are population size. The right axes of panels A and B are probability of extinction. The left axes of panels C and D are average annual harvest, and the right axes of panel C and D are coefficients of variation in annual harvest.

For the CQS, the value of H satisfying that condition is $H = 15$ (Fig. 8.7, panel A). For the CPS, the value of h satisfying that condition is $h = 0.12$ (Fig. 8.7, panel B). Furthermore, that safe limit for CPS ($h = 0.12$) yields an average annual harvest that is greater than the safe limit for CQS ($H = 15$) (compare lines with solid triangles in Figs. 8.7, panels C and D). If all that mattered were high average harvest and low risk of extinction, then CPS would be the better strategy.

However, CPS is associated with considerable fluctuations in annual harvest (open triangles in Fig. 8.7, panel D). So, if a constant harvest, year after year, was more important than high average annual harvest, then a manager might prefer CQS with $H = 15$, because that harvest has a consistent harvest each year (Fig. 8.7, panel C). Furthermore, setting h to 0.12 under CPS is associated with an *average* population size that is only ~40% of the population's carrying capacity (Fig. 8.7, panel B). If that species' ecosystem function depends on it existing at a higher density, then it may be necessary to reduce h.

The take-away from this heuristic example is that an important way to assess exploitation is to set a constraint—such as a low risk of extinction, as represented by the shaded regions of Fig. 8.7—and then compare the properties of alternative harvest strategies that meet the constraint.

Limited Knowledge and Control

The performance of a harvest strategy is influenced not only by environmental stochasticity, but also by managers' knowledge of the population's dynamics. Take a simple example: selecting a safe value of H depends on details of the solid curves in Figures 8.3 and 8.4. Those curves depend entirely on estimates of r_{max} and K. Suppose for the sake of illustration that K was overestimated by 20% and r_{max} was overestimated by 10%. That degree of overestimation is more than plausible, and the consequences for harvesting are significant (Fig. 8.8).

Managers sometimes assume that if a harvest is largely unregulated, then hunters' total collective effort will be similar from year to year, on the belief that a similar number of hunters will spend a similar number of hours hunting each year. If so, then h will be more or less the same from year to year as N fluctuates up and down. In other words, there is a common belief that largely unregulated harvests can naturally tend toward the CPS.

However, researchers have found that for ptarmigan hunters in Norway, h tends to increase as ptarmigan abundance declines, because hunters try harder when there are fewer birds. The increase in h can be impressive, with h seeming to increase from 5% to 20% as abundance declines (Erik-

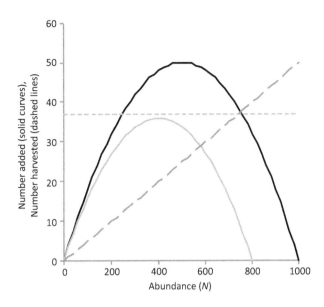

Figure 8.8   The effect of measurement error on the dynamics of an exploited population. The lighter solid curve represents a population's density-dependent dynamics ($r_{max} = 0.18$, $K = 800$). The heavier solid curve represents a manager's estimation of that same population's dynamics ($r_{max} = 0.20$, $K = 1000$). A manager might set H to 37 individuals per year under a constant quota strategy (horizontal dotted line) and think that doing so is safe for the population. In fact, the population will be overharvested and driven to extinction. Or, the manager might set h to 0.05/year (dashed line passing through origin) and expect an average harvest return of 37 per year, when the harvest return is likely to be 29 individuals, about 20% less than the manager supposed.

sen, Moa, and Nilsen 2018; Breisjøberget et al. 2018a). That is a terribly inconvenient trend, favoring overexploitation and amplifying fluctuations in abundance.

Before considering which harvest strategy is best for handling cases in which there are limits to knowledge about the population and control of hunters, let's consider two additional harvest strategies that may be thought of as modifications to CPS.

Restricted Proportion Strategy

This strategy involves harvesting a proportion (h) of the total population, but only up to a specified maximum number of harvested individuals. For example, a manager might plan to harvest 5% of the population each year, but only so long as that 5% does not exceed, let's say, 100 harvested individuals. This strategy is depicted in Figure 8.4, panel B.

If hunters' collective effort and efficiency are similar from year to year (that's a big *if*), then this harvest strategy may be approximated by imposing a daily or seasonal limit on individual hunters (e.g., two ptarmigan per day of hunting).

You can simulate these dynamics in Excel by replacing \$E\$3 in Equation 8.3 with IF(B9*H\$4 < H\$3, B9*H\$4, H\$3), which tells Excel to harvest some proportion of the population ($N_t h$), so long as that amount is less than the maximum allowable harvest (set by cell < H3 >), in which case it'll limit the harvest to that maximum.

Threshold Strategy

This strategy allows for harvesting at a certain rate (h), but only if the population's abundance exceeds some minimum threshold (Fig. 8.4, panel C). The motivation for this strategy is straightforward: when the population is low, that is the time when the population most needs a reprieve from being harvested.

You can simulate these dynamics in Excel by replacing \$E\$3 in Equation 8.3 with IF(B9 < I\$4, 0, I\$3*B9), which tells Excel to harvest nothing if abundance is lower than the threshold; otherwise, harvest a number of individuals equal to $N_t h$.

These strategies—constant quota, constant proportion, restricted proportion, and threshold—are only four of many possible harvest strategies, but they are representative of a broad swath of strategies.

Take-Aways

The performance of various strategies has been assessed under various conditions by a number of researchers. From those assessments, two notable results arise:

1. Early work suggested that the safest strategy for obtaining the greatest average annual harvest is to harvest all individuals above some set threshold and none when abundance was below that threshold (Lande, Engen, and Sæther 1995).[6] Later research indicated that this kind of threshold strategy performs adequately only if knowledge about the population's abundance and dynamics is perfectly accurate (Engen, Lande, and Sæther 1997). Taking account of uncertainties about a population's dynamics leads to thinking that an approach more like the restricted effort strategy (Fig. 8.4, panel B) is better.

2. Early work suggested that a restricted proportion strategy was best for ptarmigan harvests (Aanes et al. 2002). But later work showed that h tends to increase as N declines for ptarmigan harvests (Eriksen, Moa, and Nilsen 2018). A plausible mechanism for the

increase is hunters putting more effort into hunting when the hunting gets more difficult. In such cases, a better strategy would involve regulations designed to reduce h when N is low. This kind of strategy is also represented by the threshold strategy (Fig. 8.4, panel C).

The general lesson to emerge from those results is that the best strategy depends on whether one takes adequate account of limited knowledge about the population and limited control of the harvest.

In another study, a team of researchers played a kind of game, whereby some members of the team simulated the population dynamics of saiga antelope on the basis of published research of the species (maximum growth rates, density-dependent dynamics, and so on). Other members of the team had access to the same published information and used it to propose harvest strategies that would yield ample and reasonably consistent harvests but would not endanger the population. Then the researchers applied these proposed harvest strategies to the simulated population dynamics. This team of researchers (Milner-Gulland et al. 2001, p. 158) reported: "Many strategies were submitted; most failed to meet the criteria for success," where success was judged to be <5% chance that the population would fall below a specified threshold over a 50-year period.

The important lesson from that result is that the risk of overharvest would seem to be greatly reduced by performing formal assessments like those that we've been describing. Doing so can eliminate the strategies most likely to fail. Formal assessment is important, because intuition—even the intuition of knowledgeable people—is too often poor.

Other results from this analysis of saiga antelope indicate that when knowledge and control are limited,

- threshold strategies like that shown in Figure 8.4, panel C, are likely to perform well.
- strategies that reduce the level of harvest when r_t is declining are likely to perform well.
- frequent and adequate monitoring is essential.

While this last maxim would seem to go without saying, it is often neglected in real-world cases.

These take-away lessons are broadly applicable, but even so, it is important to know that there is no single best harvest strategy for two reasons. First, even a strategy that is potentially risky (such as CQS) can be safe if the intensity of harvest (H) is set sufficiently low. And, a strategy that is potentially safer, such as threshold strategy can be dangerous if h is set too high or the threshold is set too low. The second reason for there not being a universal best strategy is that the best strategy depends on too many contingencies that vary from case to case, such as (1) how the trade-off between large harvest return and consistent harvest return is valued, (2) how much information about the population is available, and (3) how easy it is to control the effort and efficiency of hunters from year to year as abundance changes. The importance of these contingencies reinforces the value of formal assessments tailored to particular cases.

Finally, to close the circle on those Fennoscandian populations of willow ptarmigan, it is useful to know that Aanes et al. (2002) concluded that setting h to 0.5/year is likely safe. While that number might seem high, it is important to recognize that r_{max} for ptarmigan populations is on the order of 1.5/year. Furthermore, Eriksen et al. (2018) found that h tends to increase greatly when ptarmigan density is less than ~10/km². That density is the low side of normal densities for ptarmigan in Fennoscandia. Therefore, when ptarmigan densities reach lower densities, more restrictive harvest regulations should take effect—perhaps even a threshold below which harvesting is paused until the population can recover. Obviously implementing such a harvest requires frequent and adequate monitoring.

Bio-Economics and Environmental Governance

The preceding account provides important insight about the performance of different harvest strategies, but it can fail to adequately account for the influence of some basic human behaviors that tend to affect how h changes with population abundance. To see how, it is useful (albeit cringe-worthy) to think of a wildlife population as a commodity and to think of the dashed lines in Figures 8.3 and 8.4 as gross profits that do not take account of the time, energy, or money invested to yield those profits. Human effort (which is an important determinant of h) is often more influenced by the net profit as opposed to the gross profit.

Net profit can also be influenced by demand, and demand for a commodity is sometimes greater when the commodity gets rarer (Hall, Milner-Gulland, and Courchamp 2008; Lyons and Natusch 2013). Because that tendency can increase the risk of overexploitation, it is important to anticipate the conditions when this behavior is likely to occur.

Demand can also be influenced by substitutability, which is the notion that consumers or producers seek substitutes as a product becomes rare. The process of substituting consumer preferences can alleviate demand for the rare commodity (e.g., Eriksson and Clarke 2015), and it can in a counterintuitive, but very real, way exacerbate overexploitation (Branch, Lobo, and Purcell 2013). For example, a crackdown on the illegal trade in tiger parts for traditional Chinese medicines may have the unintended consequence of intensifying the legal and illegal exploitation of lions in Africa, because many consumers of tiger parts find lion parts to be a suitable substitute (Williams 2015).

Humans also have a natural tendency to act as though today's profits are worth more than tomorrow's profits (future discounting) and to act out of "rational self-interest" as opposed to taking "collective action."[7] Both tendencies favor overexploitation. But these behaviors are only tendencies. A perennial challenge is to anticipate and mitigate the effect of such tendencies on exploitation dynamics.

A great deal of qualitative and mathematical insight exists on these topics. They are a fascinating and complex syntheses of population biology, economics, psychology, and policy. These ideas are also the purview of two interrelated fields known as bio-economics and environmental governance. For a toehold into these fields, see Clark (2010), Lemos and Agrawal (2006), and Cox, Arnold, and Tomás (2010).

STRUCTURED POPULATIONS AND SELECTIVE HARVESTS

The unstructured population models that we've been considering are useful, but structured population models (Chapter 5) are also important. From those models, you already know that changes in some vital rates of some life stages have more impact on population dynamics (λ) than changes for other stages. That circumstance is the basis for *selective harvest strategies*, which are common and important.

For example, it is common to harvest ungulate populations by focusing on males of a certain age class (as opposed to female) as a means of minimizing the impact on population dynamics. The reason this works is that many ungulates are polygynous. Because one male can mate with many females, many males can be removed from the population without reducing next year's reproduction.

This is one of many kinds of selective harvest, and structured population models are the main tool for studying these strategies. Let's begin our exploration.

Sex- and Stage-Structured Models

In Chapter 5 we learned to build projection matrices (Fig. 5.5) from life-cycle diagrams (Fig. 5.4). We'll use those ideas to build projection matrices that keep track of males and females separately. For example, many ungulate populations are well-represented by the life-cycle diagram in Figure 8.9.

To build a transition matrix from this life-cycle diagram, we'll use principles from Chapter 5. First, this projection matrix will have eight rows and eight columns, one for each of the eight sex/age classes in the diagram. We'll let the first four rows and columns represent females, and the last four rows and columns will represent males. To more easily read the projection matrix, it is useful to divide it into four regions marked by a horizontal and vertical dashed line, as shown in Figure 8.10.

Next, recall from Chapter 5 that each element ($a_{i,\,j}$) of the projection matrix is either a description of transitioning from stage j to stage i or a description of reproduction—that is, how many offspring are produced by individuals of stage j. If it is impossible for individuals of stage i to transition into stage j, then those elements ($a_{i,\,j}$) are zero, so long as these elements, $a_{i,j}$, do not represent reproduction. Determining where those zeroes belong is an easy next step in building the projection matrix. In particular,

- Males cannot become female, so the entire upper-right quarter of the matrix is filled with zeroes.
- Females cannot become males, but females can produce male offspring. So, the lower-left quarter of the matrix is filled with

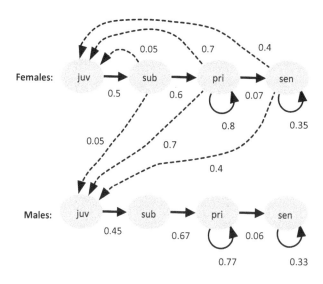

Figure 8.9 Life-cycle diagram for a creature with four age classes in each of two sexes. Notice that dotted arrows (which represent reproduction) emanate from females of reproductive age, but not from the males. For this species, none of the animals changes sex, so there are no arrows going from one sex to the other, although reproductive females produce both male and female juveniles. Otherwise, this life-cycle diagram is similar to the one shown in the lower portion of Figure 5.4.

Figure 8.10 Projection matrix corresponding to the life-cycle diagram in Figure 8.8. See main text for other details.

zeroes, except the top row of that quarter is filled with the rate at which females of different age-classes produce males.

A couple more elements will be zeroes because

· Juvenile females do not produce juveniles, and juveniles can be juveniles only for a year. So, elements $a_{1,1}$ and $a_{5,1}$ are both set to zero.

We haven't filled out all the zeroes yet. But the next easiest step is to assign non-zero values to some of the elements:

· Elements along the matrix's diagonal represent the proportion of individuals remaining in the same sex/age class for another year.
· Elements along the matrix's sub-diagonal represent the proportion of individuals transitioning to each of the various next older age classes.

The remaining elements are set to zero because individuals cannot skip age classes or transition to a younger age class.

Make sure to also observe these details about the projection matrix: I gave males slightly lower survival rates than females of the same age class, in part to help you learn which numbers go where and because it's common for males to have lower survival rates. I also assumed that the sex ratio at birth is 1:1. Because the sex ratio is 1:1, the fecundity values in rows 1 and 5 are the same. Finally, a female's total fecundity is the sum of her male and female offspring. Thus, prime-aged females of this species give birth, on average, to 1.4 offspring that survive their first year of life.

Figure 8.11 spreadsheet:

Transition matrix

	Females				Males				
	juv	sub-ad	prime	sensc	juv	sub-ad	prime	sensc	
	0	0.05	0.7	0.4	0	0	0	0	juv
	0.5	0	0	0	0	0	0	0	sub-ad
	0	0.6	0.8	0	0	0	0	0	prime
	0	0	0.07	0.35	0	0	0	0	sensc
	0	0.05	0.7	0.4	0	0	0	0	juv
	0	0	0	0	0.45	0	0	0	sub-ad
	0	0	0	0	0	0.67	0.77	0	prime
	0	0	0	0	0	0	0.06	0.33	sensc

Harvest rates, males
prime 0.9 senesc 0.9

Abundances

Year	Females juv	sub-ad	prime	sensc	Males juv	sub-ad	prime	sensc	Total	lambda	% of pop'n, male	Number harvested males prime	sensc	total
1	84.0	41.0	75.0	25.0	84.0	41.0	75.0	25.0	450.0	0.71	0.50			
2	64.6	42.0	84.6	14.0	64.6	37.8	8.5	1.3	317.3	0.95	0.35	67.5	22.5	90.0
3	66.9	32.3	92.9	10.8	66.9	29.0	3.2	0.1	302.1	1.03	0.33	29.0	83.6	112.6
4	71.0	33.5	93.7	10.3	71.0	30.1	2.2	0.0	311.7	1.02	0.33	30.1	84.3	114.4
5	71.4	35.5	95.0	10.2	71.4	31.9	2.2	0.0	317.5	1.02	0.33	31.9	85.5	117.4
6	72.3	35.7	97.3	10.2	72.3	32.1	2.3	0.0	322.3	1.02	0.33	32.1	87.6	119.7
7	74.0	36.2	99.2	10.4	74.0	32.6	2.3	0.0	328.6	1.02	0.33	32.6	89.3	121.9
8	75.4	37.0	101.1	10.6	75.4	33.3	2.4	0.0	335.2	1.02	0.33	33.3	91.0	124.3
9	76.9	37.7	103.1	10.8	96.9	33.9	2.4	0.0	341.6	1.02	0.33	33.9	92.8	126.7
10	78.3	38.4	105.1	11.0	78.3	34.4	2.5	0.0	348.3	1.02	0.33	34.6	94.6	129.2

Figure 8.11 A spreadsheet set-up for modeling selective harvest in a structured population. Important aspects of this set-up should be familiar. Specifically, rows 4 through 11 describe the projection matrix presented in Figure 8.9; cells <B17:I17> are initial population sizes; and cells <B18:I26> contain formulas for projecting abundance. This sheet also has columns to track the population's sex ratio (column L) and numbers of harvested individuals (columns M, N, and O), which are based, in part, on the harvest rates (<M13:N13>). Other details are provided in the main text.

Next year's abundance can be calculated from this matrix, a vector of eight values representing this year's abundance for each sex/age class, and the matrix multiplication that we learned in Chapter 5. Here, though, we will modify that multiplication procedure to also take account of selective harvest. Figure 8.11 shows how to perform these calculations in Excel.

For the example given in Figure 8.11, we suppose that harvesting would remove 90% of the prime-aged males and senescent males each year (i.e., $h = 0.9$/year for those sex/age classes). Formulas for projecting the abundance of unharvested sex/age classes are typed into cells <B18:I18> and then copied into the cells below. You'll determine the correct formulas with ideas you learned in Chapter 5. The key difference for this spreadsheet is the calculation of abundance for sex/age classes that are harvested. In particular, the formula for cell <H18> (prime-aged males) is

$$= \underbrace{(\text{(\$G\$10*G17)} + \text{(\$H\$10*H17)})}_{\substack{\text{Projected abundance} \\ \text{before harvest}}} * \underbrace{(1 - \overbrace{\text{\$M\$13}}^{\text{Harvest rate}})}_{\substack{\text{Proportion of individuals} \\ \text{surviving harvest}}} \tag{8.4}$$

If you compare the first part of Equation 8.4 (abundance before harvest) to the ideas from Chapter 5, you will notice that I omitted all the terms that involve multiplication by zero. I did so to make the formula more readable.

The number of prime-aged males that are harvested is found by typing Equation 8.4 into cell <M18>, except that the factor (1–M13) is replaced with (M13). The formulas for projecting abundance and harvest of senescent males are similar. This spreadsheet set-up is also easily adjusted to include harvesting any sex/age class.

After typing the formulas into row 18, copy and paste them into the cells below. Because there are no random numbers in this spreadsheet, you are able to reproduce the values in Figure 8.11 exactly as they appear there. Your instructor can provide you with an opportunity to build a spreadsheet like this to explore the effect of various selective harvests.

Consequences of Selective Harvest

A simple and important idea to glean from the preceding considerations is that males of a polygynous population can be harvested intensely without much impact on the population's growth rate or abundance. Conversely, population dynamics are far more sensitive to harvest strategies that focus on the females of a polygynous population. While those ideas include important elements of truth that are routinely implemented by managers, the fuller truth is more complicated (Milner, Nilsen, and Andreassen 2007), as illustrated by the following examples.

Example 1

Researchers analyzed harvest records of white-tailed deer for 1980–1997 from southern Ontario (Giles and Findlay 2004). The analysis indicated that efforts to harvest antlerless deer at high rates had little or no effect on deer abundance. While the regulations favored population reduction, there had likely been too few hunters to have had much impact.

This case is relevant because forest management sometimes creates habitat that can support abnormally high densities of deer. In such cases, managers sometime hope that female-focused harvest can be used to correct the imbalance. But this example shows that such an effect should not be taken for granted.

Example 2

Populations of saiga antelope were subject to severe levels of poaching beginning in the late 1980s as the Soviet Union collapsed and its rural economies deteriorated (Fig. 8.13).[8] At first, antelope were poached for meat. Before long the poaching was motivated by trade with China, where the horns of male saiga antelope are prized for traditional Chinese medicine. That economic demand fueled severe poaching of males. Because saiga antelope are polygynous, the impact of poaching was muted at first.

Figure 8.13 Saiga antelope. These are nomadic ungulates who inhabit semi-arid rangelands of central Asia. Their appearance is distinctly different from that of other species of antelope. Source: https://commons.wikimedia.org/wiki/File:Saiga_antelope_at_the_Stepnoi_Sanctuary.jpg.

Researchers studied one of these populations inhabiting a site in southwestern Russia. Throughout most of the study period (1992–2002) the percentage of the adults that were reproductive males declined from ~12% to ~5% due to poaching. (For context, that percentage in an unharvested population is typically ~20% to ~25%.) Even when the percentage dropped to ~5%, the population's reproductive rate was essentially unaffected. But then the percentage dropped further, to ~1%. At that point there were 99 females for every male—for a species that normally has one male defending a harem of 12 to 30 females. During the rut in 2000, researchers observed individual males "surrounded by large numbers of females, and dominant females were seen to be aggressively excluding subdominant females from the males." (Milner-Gulland et al. 2003, p. 135). That reversal of reproductive behavior is a likely explanation for why most first-year females did not conceive that year.

As the proportion of males dropped from ~5% to ~1%, the proportion of fecund females dropped from ~90% to less than 20% (Fig. 8.14). These changes in sex-ratio were apparently occurring across the geographic range of saiga antelope, because during this same period of time the global population of saiga antelope fell to just 5% of its former abundance.

While this case is breathtaking, it is not the only concerning case. There have been populations of caribou, elk, elephants, and primiparous moose whose selective exploitation led to female-biased sex ratios that raised concerns that the population's reproductive rate was being adversely affected (Bergerud 1974; Freddy 1987; Dobson and Poole 1998; Solberg et al. 2002).

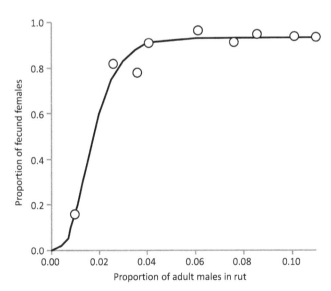

Figure 8.14 The proportion of fecund females in a population of saiga antelope (*y* axis) in relationship to the proportion of males in the population (*x* axis). Adapted from Milner-Gulland et al. (2003).

IDEAS PERTAINING TO STRUCTURED AND UNSTRUCTURED POPULATIONS

The previous sections provide basic frameworks for harvesting structured and unstructured populations. The next sections highlight ideas of general importance that extend beyond those basic frameworks.

Compensatory Mortality

Thus far, we have implicitly assumed that harvest mortality is *additive* to other sources of mortality, as opposed to *compensatory*. If harvest mortality is compensatory, it can mean either that the hunted individuals would have died even if they had not been hunted or that the removal of the hunted individuals increases the chances of survival for other individuals (due to alleviation of intraspecific competition). Knowing the degree to which harvest is compensatory is important because the impact of harvest on abundance is lessened to the degree that harvest is compensatory.

An important way to assess the degree to which harvest is compensatory is with data showing the relationship between harvest rate (h, which is usefully considered a cause-specific mortality rate) and overall survival rate (S). If the relationship is a horizontal line—that is, increasing h does not lead to a decline in S—then harvest is entirely compensatory. If S declines with increasing h and the slope of the relationship is −1, then harvest is completely additive. In other words, when harvest mortality is completely additive, there is a unit decline in survival rate for every unit increase in harvest rate. If the slope for a relationship between h and S is between 0 and −1, then harvest is *partially compensatory* (Fig. 8.15).

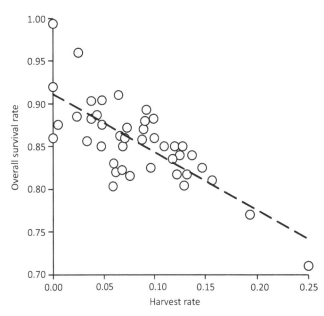

Figure 8.15 The relationship between annual harvest rate (h) and annual survival (S) for adult female elk from 43 elk populations in the western United States. The value of R^2 for this relationship is 0.47, which indicates that about half of the variation in survival rate is attributable to variation in harvest rate. The estimated slope of the relationship is −0.63, and the 95% confidence intervals for the slope do not include 0 or 1. Harvest mortality is partially compensatory for these populations. Adapted from Brodie et al. (2013).

Harvest mortality may also be compensated by an increase in reproduction. Compensatory reproduction is a reason why it is difficult (essentially futile) to reduce the abundance of some species through exploitation. Coyotes and black jackals are examples (Knowlton, Gese, and Jaeger 1999; Minnie, Gaylard, and Kerley 2016). Finally, it is also possible for harvest to be super-additive such that a one-unit increase in harvest rate is accompanied by a more than one-unit decline in population growth rate (Vucetich, Smith, and Stahler 2005).

The nature of harvest mortality has been studied intensively and extensively for many species of Anatidae (family that includes ducks and geese). The take-away from these studies is that the degree to which harvest mortality is compensatory varies among species, likely varies among populations of a given species, and likely varies within a population over time (Cooch et al. 2014; Pöysä et al. 2004, 2013). That variability also seems to apply to Phasianidae (family that includes grouse and ptarmigan). The safe assumption for any particular case would be that mortality is additive unless there is evidence to the contrary.

That being said, there is a reasonable sense that harvest mortality tends to be more compensatory:

· for species with high fecundity, low survival, and high capacity for dispersal.
· for populations that are at or near carrying capacity.
· when harvest rates are low.
· when harvest occurs at a time of year that precedes seasonal periods of high natural mortality (e.g., fall harvest for species where most non-harvest mortality occurs in winter). (Sandercock et al. 2011)

The tendency for harvest mortality to be increasingly additive when abundance is low is important to consider when hunting populations in decline, even if hunting is not thought to have been a cause of the decline. That prospect is the reason for concern over harvesting willow ptarmigan in Fennoscandia.

Impacts beyond Demography

Exploitation reduces population abundance (Fig. 8.2) in large part by reducing survival rate. Reductions in survival can be dramatic. For example, for bighorn sheep in Alberta, Canada, the probability that a four-year-old male will survive to become eight years old is about 58% in a non-hunted population, but only 27% in a trophy-hunted population. Similar reductions in survival rate have been observed for male red deer and female moose in Norway. More generally, most mortality in most of the world's ungulate populations is caused by hunting (Festa-Bianchet 2003).

Large reductions in survival can lead to broad changes in life history. To see how, appreciate that life history traits are often traded off. For example, individuals who invest more effort reproducing earlier in life tend to die younger.[9] Furthermore, an ungulate's reproductive success involves some luck, as weather conditions have an important influence on the survival of offspring from conception through to the first year of life. That stochastic element means that longevity confers a fitness advantage by providing more chances to successfully reproduce. But the price to pay for longevity is often a reduction in an individual's annual effort toward reproduction. This reduced effort may be observed in the form of smaller litter sizes, lower birth weights, lower juvenile survival, and delaying the age of first reproduction. The relevance of those life history dynamics for heavily hunted populations is that reductions in the longevity of females (due to hunting) can lead to increased annual effort in reproduction, often observed in the form of giving birth to larger offspring and larger litters and reproducing for the first time at a younger age (Festa-Bianchet 2003). Similar impacts have been reported in response to the exploitation of fish populations (Heino, Díaz Pauli, and Dieckmann 2015).

Exploitation can also lead to population-level changes in anatomical phenotypes. One of the best documented cases involves intense poaching of elephants in Mozambique to obtain ivory to finance a civil war from the late 1970s to early 1990s. While all males possess tusks, females vary in terms of possessing tusks. Tuskless females were far less likely to be poached. Before the war, about 19% of females were tuskless, but among females born after the war 33% were tuskless. There is enough evidence from this case to further indicate that the change was evolutionary—that is, the change in frequency of the phenotype was brought about by a change in the frequency of genotypes (Campbell-Staton et al. 2021).

Selective exploitation has also been implicated in changes in the size of horns and antlers for several populations of ungulates and for changes in the body size of fish. Some of these cases have been difficult to evaluate due to challenges in determining whether the changes are due to exploitation or some undetected environmental change. In addition, when exploitation is reasonably implicated, there have been challenges in determining the extent to which the changes in phenotype are due to changes in gene frequencies or phenotypic plasticity (Heino, Díaz Pauli, and Dieckmann 2015; Festa-Bianchet and Mysterud 2018).

The impacts of exploitation routinely extend beyond survival and abundance.

Resiliency to Overexploitation

If a population is inadvertently overexploited, it might seem straightforward to stop the harvest and let the population rebound. A technical term for this rebounding is *resiliency*. The quicker the rebound, the more resilient the population. The concern is that those involved with a harvest (e.g., scientists, managers, and harvesters) are not always good at anticipating the limits to nature's resiliency.

The tendency for some species to be more resilient than others is sometimes explained by differences in life histories. Some differences seem obvious. A healthy population of willow ptarmigan can more than double its abundance in a single year, but a healthy population of tigers cannot grow any faster, over the long term, than ~3%–5% per year (Miquelle et al. 2015). It's easy to expect that an otherwise healthy population of tigers would be less resilient to overharvest than an otherwise healthy population of willow ptarmigan.

Less obvious differences in life history can also explain important differences in resiliency, such as appears to be the case for Mysticetes and Odontocetes (Wade, Reeves, and Mesnick 2012). Mysticetes are the baleen whales and include blue whales and fin whales. Odontocetes are the toothed whales and include sperm whales, belugas, and dolphins. Across the wide range of life histories found among animals, Mysticetes and Odontocetes seem to be two subtly different versions of the same kinds of creatures— large-bodied, long-lived, and slow at reproducing. Nevertheless, they differ enough to have different resiliencies to overexploitation. Many populations of Mysticetes and Odontocetes have been overharvested to very low levels, sometimes less than 10% of their original abundance. From the 1980s onward, there has been a tendency to better protect Mysticetes and Odontocetes. As a result, many populations of Mysticetes have shown significant signs of recovery. But Odontocete populations seem not to have shown the same propensity for recovery. The difference in resiliency is explained by what might seem to be small differences in reproductive life history (Lilley, Smith, and Botero-Acosta 2017). In particular, the age at first reproduction tends to be a few years older for Odontocetes (10–12 years) than Mysticetes (5–8 years). And the interval between birthing for an individual female is typically longer, on average, for Odontocetes (1–14 years) than for Mysticetes (1–5 years).

Tigers, leopards, and cougars would also seem to be relatively minor variations on the same life history theme of being a large felid. Yet, subtle differences in life history explain why tigers are likely less resilient to over-harvest than leopards and cougars (Chapron et al. 2008). Similarly subtle, yet important, differences in life history make forest elephants far less resilient to overexploitation than savanna elephants (Turkalo, Wrege, and Wittemyer 2017). An important take-away from these examples is that resiliency (or lack thereof) is not always easy to predict in advance of careful analysis like that presented by Wade, Reeves, and Mesnick (2012) and Chapron et al. (2008).

In addition to life history traits, resiliency can also be limited by the intricate interconnections of an ecological community. For example, cod populations collapsed in the northwest Atlantic Ocean and the Baltic Sea in the 1990s. The collapses were rapid and severe, with abundance dropping to less than 15% of previous population highs. Commercial fishing was halted, causing great economic hardship for those involved with cod fisheries. In the many years since harvesting ceased, these cod populations have shown little or no recovery. The lack of resiliency is not fully or certainly understood, but an important possibility is that abundant cod populations are supported by eating lots of small (young) sprat, which are quickly replenished by larger (older) sprat with high fecundity. When cod are overharvested, small sprat experience less predation and consequently higher survival rates. The larger number of surviving sprat compete with each other, resulting in stunted growth. Stunted growth is consequential, because small sprat are not fecund enough to support an abundant population of cod.

In other words, an abundant cod population maintains the size-structure of sprat in a way that is favorable to supporting abundant cod. But when cod are overharvested, the size-structure of the sprat population changes in a way that is unfavorable for cod and is not readily changed back. Those dynamics are consistent with cod not recovering from overexploitation after the exploitation stopped (Van Leeuwen, De Doos, and Persson 2008, McCann 2011; see also Swain and Benoît 2015).

Similar ecological interactions can occur with bass and sunfish in freshwater lakes (Walters and Kitchell 2001). Populations of bass (predator) and sunfish (prey) exhibit size-dependent predation. Larger-bodied adult bass prey on sunfish, the largest of which are smaller than adult bass. But the largest sunfish are capable of preying on juvenile (smaller-bodied) bass.

When larger adult bass are abundant, they limit the abundance of sunfish, which greatly limits the degree to which juvenile (smaller) bass are eaten by or compete with sunfish. As a result of those interactions, there is a tendency for bass to stay abundant and sunfish to remain less abundant. But if bass abundance is reduced—say, through harvest—then the system changes in a way that is not easily changed back. Specifically, when adult bass are depleted, sunfish populations increase. The increased abundance of sunfish leads to more predation on and competition with juvenile bass. With juvenile bass limited in that way, the bass population tends not to recover.

For a third example of limited resiliency to overexploitation, consider Yellowstone National Park from which wolves were extirpated during the early twentieth century due to overexploitation. Afterward, elk became overabundant, which led to over-browsing in riparian communities of willow. The loss of willow led to the loss of beaver from Yellowstone's streams. With the reintroduction of wolves to Yellowstone in the mid-1990s, those changes were expected to reverse themselves. While many ecosystem conditions changed with the restoration of wolves (Ripple et al. 2014), riparian willow communities were not restored as expected. Field research indicates that the lack of resiliency is tied to a broken feedback loop, whereby willow reestablishment requires hydrological conditions associated with rivers that are dammed by beaver, and the reestablishment of beaver requires the presence of a more intact riparian community of willow (Marshall, Hobbs, and Cooper 2013).

The lesson of these examples is simple: it is unwise to take for granted a population's resiliency to overexploitation, because resiliency is difficult to predict, and the damage caused by lack of resiliency can be significant.

Risk of Overexploitation

Inasmuch as the consequences of overexploitation are adverse, there is value in better understanding the risk of overexploitation. That risk would seem to increase under the following conditions:

- Hunters are motivated (and able) to exploit more than is sustainable and their behavior is not readily controlled by managers. This characterizes some instances of poaching and bushmeat harvesting. Responses to such cases include intensifying efforts to control hunters' behavior (law enforcement) or trying to better understand and alter hunters' motivation (e.g., Muth and Bowe 1998; Kahler and Gore 2012; Montgomery 2020). Such efforts are often difficult to make effective.

- The influence of sound scientific advice about exploitation is diminished by undue political pressures, often tied to stakeholders' preference to harvest at a rate that ends up being unsustainable. When such occasions arise, it is not uncommon for opponents of stricter regulations to argue for waiting just a bit longer before enacting stricter regulations—in hopes that the population will rebound without stricter regulations. Such circumstances contributed, for example, to the collapse of cod fisheries in the northwest Atlantic Ocean (Harris 2013).
- Managers are overconfident about some of the assumptions discussed in the earlier section on "Implicit Assumptions." That overconfidence can be manifest as undue belief that the assumptions are met or undue belief about the consequences of violating those assumptions (see also Fig. 8.8).

Overconfidence

There is solid scientific reason to be concerned about the overconfidence of experts. To see how, consider a common way for psychologists to study overconfidence: imagine asking hundreds of financial experts the percentage by which they think the S&P 500 market will grow or contract over the next year. You'd get a range of answers such as 2% growth, 3% contraction, and so on. After soliciting these predictions, wait to see how each prediction compares with the actual change in the stock market. Unsurprisingly, many experts will get the wrong answer and many will be close to the right answer. That result would *not* be the interesting part. The interesting result comes from also having asked these experts to give a range of values for which they are 80% confident that the market's actual change will lie within. For example, an expert might say they are 80% confident that the market's change over the next year will be between −5% and +1%. That expert would then be "surprised" if the actual change fell outside that range.

Imagine asking hundreds of experts to provide 80% confidence intervals for many predictions on their topics of expertise. If they have an accurate understanding of the limits of their expertise, then they will be surprised about 20% ($= 1-0.80$) of the time. For the experiment that I am describing with the financial experts, they were surprised about 67% of the time— which is to say, most of the time. That is a stunning level of overconfidence. Experiments like this have been conducted for a wide range of expertise types (medical doctors, law enforcement officers, engineers, and more). The overall result is that experts—regardless of their field of expertise—are prone to high levels of overconfidence about their judgments.

More is known about why overconfidence occurs, the kinds of conditions that favor overconfidence, and some remedies. That additional knowledge explains why expert intuition is sometimes marvelous and sometimes deeply flawed. There is every reason to think that this psychology applies to experts in the field of natural resources and conservation.

Going further on this topic is beyond our scope, but you can learn more by searching the internet for the phrase "psychology of overconfidence" or by reading Kahneman (2011). For now, the take-away message is that natural resource managers should be extremely cautious about being overconfident.

Purpose Can Determine Success and Failure

Population collapse is not the only basis for judging whether exploitation has been a success or failure. To see how, consider the Northern Range elk herd that spends part of the year in Yellowstone National Park and part of the year outside the park on lands that are also heavily grazed by cattle. During the 1970s, before wolf predation had been restored to the region, elk abundance was high, and livestock owners objected to elk eating so much forage that they believed was intended for cattle. In response to the concern, wildlife managers from the state of Montana began implementing harvest regulations whose explicit purpose was to reduce elk abundance. The regulations included higher rates of harvest of females during late winter when harvest mortality is most likely to be additive. These harvest regulations were known as the "late hunt." In spite of the exploitation, elk abundance remained high for many years. Then wolves were introduced to Yellowstone in the mid-1990s. In the decade that followed, wolf abundance grew but remained too low to be of significant influence on elk abundance. But that decade following wolf reintroduction was also associated with a multi-year drought (which reduced elk forage) and managers' decision to increase the rate of harvest. As a result, elk declined during that decade. From that point forward, as elk abundance continued to decline, managers reduced the harvest rate and eventually canceled the late hunt entirely. The harvest's explicit goal to reduce elk abundance had been realized. Success, right? Well, maybe. The concern is that many hunters grew angry about the lost opportunities to hunt elk. Hunters and some managers mistakenly blamed wolves for diminished opportunity to hunt, when in fact the elk decline in the decade that followed wolf reintroduction was almost certainly caused by harvest and drought (Vucetich, Smith, and Stahler 2005). It might be said that the anger was born from having forgotten the purpose of the hunt—namely, to reduce elk abundance. If, instead, the purpose of the hunt had been to maximize harvest return over the long term, then a different set of regulations would likely have been called for.

Because exploitation is fundamentally a goal-oriented endeavor, it can succeed or fail with respect to its goal or purpose. The study of project failures in conservation is rare. This is unfortunate because institutions that embrace the study of failed cases are in the best position to improve. In one of the few studies of project failures in conservation, researchers found that the most common kind of failure involved attitudes of and relationships among the humans involved in the conservation project (Catalano et al. 2019). The risk of such failures can be greatly reduced by developing plans that include the transparent documentation of (Vucetich et al. 2017):

1. the goal or purpose (of the exploitation), expressed in adequately precise terms.
2. the rationale for why the goal or purpose is appropriate with respect to the stakeholders' interests.
3. the reason to think the goal will be realized with the methods to be used.
4. objective, measurable criteria for determining if the goal has been successfully achieved.

Those four elements can also be easily remembered by the questions What? Why? How? In other words, what is the goal? Why is the goal appropriate? How will the goal be accomplished, and how will success be judged? Following those principles is sometimes trickier than is at first apparent, as illustrated by this next example.

In Michigan, plans to hunt wolves were developed and implemented in 2013. Official documents indicated that the purpose of the hunt was to reduce wolves' threat to human safety and livestock (Mason et al. 2013). There is no doubt that keeping humans and livestock appropriately safe are worthwhile purposes. But the harvest plans did not acknowledge that wolves are not a threat to human safety except in the rarest and most unusual circumstances or that rates of livestock depredation were very low (about a dozen depredated cattle per year). Furthermore, recreational hunting is not a sensible way to protect humans or livestock from those circumstances. When livestock or a human has been threatened by a wolf, effective responses require an immediate response focused on the offending wolf or wolves. Recreational hunting does not lend itself to such a targeted response.

To bring about greater safety to humans and livestock, the plan called for killing no more than 43 wolves at a time when there were an estimated

625 wolves.[10] Analysis indicated that this level of harvest would be very unlikely to reduce the number of livestock depredations or make humans
any safer (Vucetich et al. 2017). The plan made no mention for how it would
determine whether the harvest was successful in realizing its purposes.
Those perspectives raise significant doubt about whether questions 1
through 4, listed previously, were adequately answered. Furthermore, the
real (yet unstated) purpose of the harvest was likely to provide citizens who
hate wolves with a chance to legally kill a wolf. That purpose would have
been difficult to justify and likely rejected by most Michiganders. In any
case, the Michigan Department of Natural Resources asserted that hunt a
success (Klug 2014) without providing any evidence that it achieved its
stated purpose.

Different people will see these two cases (Michigan wolves and Montana elk) differently. Regardless, the point is that exploitation can be judged
as a success or failure with respect to its purpose. Furthermore, an important means of assessing purpose is whether questions 1 through 4 are adequately answered. Those four questions are also a robust, flexible, and
straightforward framework for integrating the scientific, social, and ethical dimensions of a plan. Finally, to recall the other points of this section,
population collapse is not the only way that exploitation can fail, failure
often rises from attitudes of and relationships among stakeholders, and institutions that embrace the careful assessment of failures are in the best
position to improve.

The Ethics of Exploitation

If you believe there are some occasions for which it is likely acceptable to
kill a sentient animal, then the ethics of exploitation can be complicated.
Furthermore, there are several distinct and major ethical elements of
exploitation.

The first is that the method of killing matters. Some methods involve
more suffering and more non-target deaths. For example, poison and traps
are less discriminating than rifles. It is too simple to say that the methods
involving the least suffering and fewest non-targets should be used. The
harder question, by far, is, Is a particular method sufficiently humane and
targeted? If not, and if the exploitation is not vital to human well-being,
then perhaps that method should not be allowed.

Methods that are most humane and least likely to result in wounded animals getting away are sometimes less likely to be associated with "fair
chase," which is valued by some hunters. For example, hunting ungulates
with dogs is typically not allowed in the United States. But moose hunters
in Sweden are required to use dogs to eliminate the chance that a wounded

animal escapes. In other words, there is a trade-off between fair chase (which disfavors the use of dogs) and minimizing wasted suffering (which favors the use of dogs).

The second ethical issue concerns what counts as a good reason to kill. This question is often difficult to answer, but it must be answered. Some of the difficulty pertains to the benefits of exploitation being so varied from case to case. For example, the benefit of bushmeat to a rural person in a low-income nation is not the same as the benefit of a grizzly bear skin to a trophy hunter. In any case, the only formal, rational method for assessing the question is to perform argument analysis like that demonstrated in Chapter 4. For examples of that technique applied to killing wildlife, see Vucetich et al. (2019) and Coals et al. (2019). Another aid to such analysis is the four questions raised in the section "Purpose Can Determine Success and Failure."

Finally, there is also a modest body of scientific literature on people's attitudes toward wildlife exploitation and why they hold those attitudes. A basic lesson from that literature is that most citizens are non-hunters and that many non-hunters support hunting, but that support is contingent on the reason for hunting and the method of hunting. For toeholds into this literature, see Blascovich and Metcalf (2019) and Mudumba et al. (2022) and references therein.

BEST PRACTICES AND COMMON PRACTICES, REVISITED

Toward the beginning of this chapter, I mentioned that it is common for there to be a significant gap between best practices and common practices with respect to the design and implementation of exploitation. Best practices include analyses along the lines described in this chapter. Yet one of the most common ways to implement recreational exploitation in high-income nations is to adjust harvest regulations (making them more or less restrictive) for an upcoming harvest according to the population's response to recent harvests (Sutherland 2001). And it is common to make those adjustments without giving due attention to the four questions presented in the section on "Purpose Determines Success and Failure."

The kinds of harvest regulation that managers can adjust include number of hunters, length of season, methods of hunting (e.g., use of tree stands or bait piles), individual bag limits, quotas, and selectivity of harvest. While such adjustments can be effective, there is typically only a rough sense for how changes in regulation actually translate into changes in harvest rate (h). And intuition is not always an adequate guide, as illustrated by the example of white-tailed deer harvest in Ontario. In that case, increasing the number of hunters permitted to take antlerless deer resulted in increased h (as expected), but only up to the point that 40% of hunters had such a permit. Further increases in permitting had no effect on h. Furthermore, that information was not available to managers until many years after the fact.

Considerable discernment is required to judge this gap between best practices and common practices because while overexploitation is a grave threat to conservation, many harvests are implemented in a reliably sustainable manner. Furthermore, many harvests that were thought to be reliably sustainable (at least by some) ended up being otherwise.

CHAPTER 9

Predation

Tyger Tyger, burning bright,
In the forests of the night;
What immortal hand or eye,
Could frame thy fearful symmetry?

And what shoulder, & what art,
Could twist the sinews of thy heart?
And when thy heart began to beat.
What dread hand? & what dread feet?

When the stars threw down their spears
And water'd heaven with their tears:
Did he smile his work to see?
Did he who made the Lamb make thee?

Excerpted from William Blake's "The Tyger"

The density-dependent dynamics that we studied in Chapter 3 offer an explanation for why populations fluctuate without growing to infinity or tending to go extinct. Those dynamics can represent cases in which a population's food supply diminishes on a *per capita basis* as abundance increases. That circumstance implies that the *total supply* of food is a constant. In other words, the food is renewed at the same rate that it is consumed. In this chapter, we will account for cases in which consumption by predators causes a decline in the abundance of the resource (prey), and the predator population is subsequently affected by any such decline. In other words, we'll be looking simultaneously at two populations that are dynamically linked to one another by predation. We'll be studying how predation affects per capita growth rate of populations when predation is a source of food for predators and a source of mortality for prey.

Knowledge about the population ecology of predation developed slowly. Worthwhile insight follows from a sketch of that history. Populations' propensity for exponential growth (Chapter 2) was first presented in 1798 by Thomas Malthus as part of an argument that the long-term exponential growth of human populations would exceed the linear rate of growth for humans' capacity to produce food. The idea that food limitation can lead populations to exhibit density-dependent growth (Chapter 3) appeared in an 1838 report by Pierre François Verhulst, who was aiming to discover mathematical laws that explain population dynamics in much the same way that mathematical laws explain physical phenomena, such as Newton's laws of motion. Those early ideas are antecedents to discoveries about the population ecology of predation made by Alfred Lotka and Vito Volterra in 1925 and 1926, respectively.

From those first bits of history, you can appreciate that the ideas we've been touring in this book are a compression of thought developed over the past 225 years. Recognizing science as a history of ideas discovered by real people allows one to ask, What motivated those discoveries? Understanding those motivations is often an aid to understanding the knowledge itself. That is very much the case for predation.

The pursuit of predation knowledge has two kinds of motivation—basic science and applied science. The applied motivations may be familiar to you already. They involve an ability to answer questions such as how does predation affect prey populations? If there are fewer predators, for example, will there be more prey? If so and if the prey is a pest (such as an agricultural pest), can predation be added to an ecosystem to reduce prey abundance? Conversely, if the prey is a valued resource (such as an exploited fish population), can predators (such as seals) be killed to increase prey abundance?

The basic science motivations are related, but more subtle and perhaps less familiar to you. They arise from an observation so obvious as to be easily overlooked: life on Earth is diverse and abundant with myriad species wherever one looks. How do all these forms of life coexist, given how common predation is and the sketchy but persistent intuition that predation would seem to risk driving prey populations to extinction?

Minding those motivations as we proceed will add considerably to your understanding.

LOTKA-VOLTERRA EQUATIONS

After returning from service in World War I, Umberto D'Ancona found himself working for the Italian government as a fish biologist when he found catch data from several fishing ports along the northern Adriatic (Fig. 9.1). The data showed the annual proportion of total catch that was "selachian," an antiquated term referring to sharks, rays, and skates. These predators were tracked because they were considered unpalatable and therefore of less value to the fishery.

Prior to World War I, selachians represented about 12% of the catch. With the onset of war, fishing was severely curtailed, and the portion of the catch that was selachian increased threefold. With the war's end, intense fishing resumed, and the share of selachians fell to prewar values. D'Ancona believed that intense fishing had tipped the balance between predator and prey in the years prior to the war, that balance was restored during the war, and that it then tilted again afterward.

For all of D'Ancona's excitement about that data, he was short on the details that might explain it. Conveniently, D'Ancona was courting the daughter of a celebrated mathematician, Vito Volterra. D'Ancona piqued

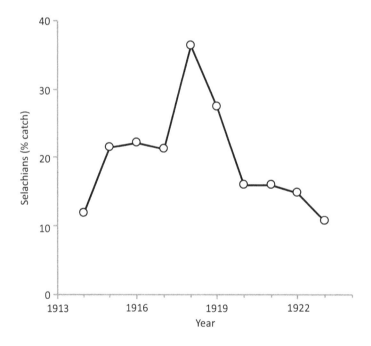

Figure 9.1 The rise and fall of selachians caught from the Adriatic Sea in the years surrounding World War I, 1914–1918. Vito Volterra was inspired to find equations that could explain these fluctuations. Data from Braun (1983).

Volterra's interest, who dreamed of discovering the laws of nature that governed interactions between populations of predator and prey.

Let's trace the thinking that supports Volterra's formulation for a pair of equations representing two populations that are dynamically linked to one another by predation. The first step is to recall from Chapter 2 that a population growing without limits grows exponentially and according to this equation:

$$N_t = N_o e^{rt}. \tag{9.1}$$

Of that equation we can ask, at what rate does N change given a small change in t? The answer to that question is another equation called the derivative of Equation 9.1, which looks like this

$$dN/dt = rN. \tag{9.2}$$

The meaning of that equation in English is important and easy: for a small change in time (dt), abundance will change (dN) by this amount, rN. Equation 9.2 has the same meaning as

$$N_{t+1} - N_t = N_t r_t, \tag{9.3}$$

which is a simple rearrangement of Equations 2.1 and 2.2.

Equation 9.2 is a differential equation, in which time is treated as a continuous variable; Equation 9.3 is a difference equation, in which time is treated as a discrete variable. Otherwise, the two equations say the same thing. I present Equation 9.2 because Volterra's line of thinking is easier to follow using the notation associated with differential equations.

Now, recall the strategy we used to derive an equation for density-dependent growth back in Chapter 3. We conceived of an idea (i.e., r should decline with increasing N), we represented that idea as a simple graph (in Figure 3.3, panel B), and then we figured a way to modify Equation 9.3 to represent those dynamics.

Volterra took a similar approach. In particular, he started with Equation 9.2 to represent a population growing without ecological limitation. Then he thought how such a population might be limited by predation and modified Equation 9.2 accordingly.

That imaginative process can be represented with a hypothetical population of prey—let's say lemmings. Suppose the lemmings live on an arctic tundra with as much food—willows, sedges, grasses, and rushes—as they could ever want. They will live full lives, give birth to many offspring, and over time their abundance will increase. Those ideas are symbolized by rN, where N stands for the number of lemmings at any particular time, and r is the rate by which lemming abundance increases. To help our imaginations, let's replace N and r with numbers. Suppose there were 100 lemmings. If they grew at a rate of 0.05, then soon we'd have 5 new lemmings ($100 \times 0.05 = 5$) in addition to the initial 100.

You already know that populations exhibiting exponential growth at rates (r) realizable by many animal populations quickly grow to wildly high abundances. But suppose these lemmings also live in the presence of predators—imagine stoats, a kind of weasel. The stoats kill lemmings at some specific rate or frequency. How frequently depends on how quickly each stoat finds their next lemming. More lemmings will mean more frequent killing—mainly because more lemmings make it is easier for a stoat to find a lemming. Fewer lemmings would mean less frequent killing. Those ideas may be expressed mathematically as aN, where a is a rate of encountering and killing prey. The variable a is sometimes referred to as the *attack rate*. If we have 100 prey, and if a is 0.02, then two lemmings will be lost to each stoat in the population ($0.02 \times 100 = 2$). In other words, the product, aN, is the *per capita kill rate*, which is the number of prey killed by each predator during the given amount of time.[1] The total number of lemmings lost to predation is that quantity multiplied by the total number of stoats in the population, which is aNP, where P stands for the number of predators. If there are three stoats, and each kills two lemmings, then the total loss of lemmings would be six.

Thus far, we have two ideas, rN and aNP, where rN is the growth that would occur in the absence of predation and aNP is the loss due to predation. In Chapter 8, we accounted for losses due to harvest by subtracting the loss from the gain. We'll do the same here, and the result is an equation for prey dynamics:

$$dN/dt = rN - aNP. \qquad (9.4)$$

If the number of predators (P) never changed, we'd be done. But that's not the case. Instead, we'll expect the number of predators (P) to change depending on the abundance of prey (N). Or more precisely, the change in the number of predators from one time period to the next should depend on how much prey each predator has eaten. If so, we might start like this:

$$dP/dt = f(aN).$$

Recall from an introductory algebra class that the notation $f()$ means "is a function of." In this equation, the meaning is that the change in predator abundance over some small change in time (dP/dt) is some function of aN, where aN is the amount of prey each predator ate during that period of time. In other words, we need to find a way to convert aN (prey consumed per predator) into population growth for the predator (dP/dt).

Volterra's first thought about this function, $f(aN)$, is that there would be some efficiency (e) by which predators convert consumed prey into their own flesh. Think of efficiency as answering the question, How many prey do predators need to eat to survive and reproduce at a specified rate? Efficiency would be different for different kinds of predators and prey. For example, e would be a smaller value for predators that eat prey smaller than themselves (whales eating krill) and larger for predators eating larger prey (wolves eating bison). Clearly, one bison means more to a wolf pack, than one krill does to a whale. In any case, this conversion process from prey biomass to predator survival and reproduction can be represented as

$$dP/dt = eaNP.$$

Think of this equation as $r_{pred}P$, where r_{pred} is the per capita growth rate of the predator population, calculated as eaN, or the number of prey eaten per predator (aN) times the efficiency (e).

Volterra thought another adjustment would be in order. If a predator population was starved, it wouldn't decline to zero immediately. Rather it would die at some rate, say m.[2] To account for that idea, he further modified the predator equation to be

$$dP/dt = eaNP - mP. \qquad (9.5)$$

That's it! Equations 9.4 and 9.5 are referred to as the Lotka-Volterra equations for predator-prey dynamics.[3] The entire set of ideas is summarized in Figure 9.2.

To better understand how these equations work, return to the lemmings and stoats. Recall that we started with 100 lemmings to which 5 new lemmings were added ($rN = 0.05 \times 100 = 5$) and 6 were lost to predation ($aNP = 0.02 \times 100 \times 3 = 6$), leaving 99 lemmings. That is not a final answer, merely the change in lemming abundance over a short period of time. Suppose that e and m from Equation 9.5 are set to 0.39 and 0.10, respectively. If so, then the number of stoats would change by two. (You can do the arithmetic on your own.) Where we had three stoats before, we would now have five stoats.

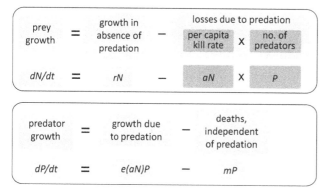

Figure 9.2 Lotka-Volterra equations and explanatory diagrams. The lower lines of each box are the equations. The upper lines of each box explain each term of the equations. Notice that the equations are linked by the quantity aN, which represents the per capita kill rate.

The changes depend not only on those rates (i.e., the lower-case letters in Equations 9.4 and 9.5) but also on the abundance of predators (P) and prey (N), which are also changing over time. With 99 lemmings and 5 stoats, the equations tell us that the lemmings will soon decline to 94, and the stoats will increase to 9. The lemmings decline, and the stoats increase. If that trend continues, the lemmings will go extinct, at which time the stoats would be committed to the same fate. But the rates of predation will slow as the lemmings decline, at which point the stoats will be eating less. Perhaps lemmings will persist and with them the stoats, but only if the rates are balanced just right. It's far from obvious that it would turn out this way.[4]

The Celebrated Result

No human can just look at Equations 9.4 and 9.5 and know the outcome. The equations need to be analyzed. When Volterra analyzed these equations, he found that the abundance of predator and prey fluctuated over time, in perpetuity (Fig. 9.3). At first, the predator declines, while prey rise. Then comes a shift, and both populations rise for a short spell. Another shift, and prey decline while predators continue to rise. Before long, the predator population reverses course and starts to decline. Now both populations are on the decline—but only for a while, after which the prey begin to rise. The sinusoid is complete and begins afresh.

Throughout this book, we've been motivated by the question, What causes a population to fluctuate? The answers we've gathered include the following:

· Environmental stochasticity buffets populations up and down away from their equilibria (K), while density-dependent forces tend to press populations back toward their K (Chapter 3).

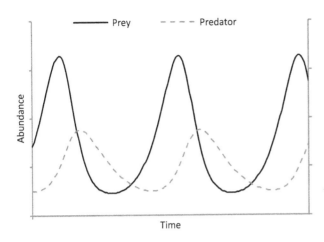

Figure 9.3 Graphical result of the Lotka-Volterra equations. The equations predict that the abundances of predator and prey fluctuate over time, exhibiting regular population cycles.

- Time delays in density dependence can lead to fluctuations in abundance even in the absence of environmental stochasticity (Eq. 3.5; Fig. 3.11).
- Environmental stochasticity can affect individuals of some age classes more than others. This leads to changes in age structure, which in turn can generate fluctuations in per capita growth rate (Fig. 5.3).

Now we can add another important answer to that question:

- Populations fluctuate when they are tightly bound to one another through predation—even in the absence of environmental stochasticity.

By "tightly bound," I mean that the prey's primary source of mortality is the predator, and the predator's primary source of food is the prey.

This chapter opened with poetry to remind us that an adequate understanding of predation transcends any technical description of equations. Yet the equations most certainly provide distinctive inspiration for poetic understanding. Volterra assembled an inert string of symbols and found two populations heaving and sighing in synchrony, rising and falling like waves on an ocean. He discovered a rhythm in an immortal sequence of births and deaths. Aided by the precision of math, one can sense a pulse rise from the undulating balance between dewy newborns and bloodied corpses. Extinction contraposing boundless proliferation.

Volterra was delighted to know that his equations bore resemblance to Figure 9.1. But he also knew that he hadn't proved that his equations were the best explanation for those fluctuations. And he knew that his equations—while complex—were also very simple compared to all that happens between predators and prey. Nevertheless, it seemed plausible at the time (and more so today) that the essential feature of any explanation of predation is a pair of equations linked by the per capita kill rate (aN), which represents the rate at which predators acquire food as a fundamental source of mortality for prey.

Coexistence Is Delicate

Aside from Figure 9.3, the Lotka-Volterra equations lead to another result—less celebrated, but no less important. This result can be apprehended by first appreciating that tracking the number of predator and prey over time is analogous to tracking the money in two linked bank accounts. The balance in either bank account is dependent on the interest rates earned, the interest rates charged, and whether any of the charges are deposited into the other account. What happens to the bank accounts over the long term depends on the details of those rates.

On the other side of this analogy, the Lotka-Volterra equations have rates of increase (first terms in Equations 9.4 and 9.5) and rates of loss (second terms in each equation), for which some of the losses to one population are given to the other population. As with the linked bank accounts, what happens to the populations over the long term depends on the details of those rates.

Volterra's equations involve four rates. One rate is associated with the intrinsic nature of the prey—how quickly it can grow when it has all the food it wants, r. Another rate is tied to the predator—how quickly it dies in the complete absence of food, m. The other two rates describe predatory interactions between the two populations: the attack rate, a, and the efficiency with which predators convert captured prey into their own kind, e. The four rates pull and push on the populations, each in its own direction. The force behind each rate changes continuously as the numbers of predator and prey change. Just as one population is about to rocket toward infinity, the other population takes hold and the pattern reverses. What we have is the most abstract, rarified, and bloodiest game of teeter-totter imaginable.

That exposition on rates is a set-up for this important but simple result: change any of those four rates, and the dynamics will change.* Change any rate by *too much*, and one or both populations will go extinct. More precisely, predator and prey coexist for only a narrow range of parameter set-

---

*The amplitude and frequency of the cycles are, for example, a function of the four rates.

tings for r, a, e, and m. Most parameter combinations result in one or both populations going extinct.

Taking account of environmental stochasticity would only exacerbate the risk of extinction. A few bad years in the environment timed with one of the populations being at the low part of its cycle and blink: it's gone.

According to the Lotka-Volterra equations, predator-prey dynamics seem delicate to the point of being a concern. On the one hand, the reasoning that gives rise to the math is compelling, even though no one thinks it's a perfect representation of nature. And that math suggests that coexistence between predator and prey is, broadly speaking, unlikely. On the other hand, many predator populations seem to coexist for long periods of time with the prey populations upon which they depend.

The stakes are, in fact, a bit higher: most species eat other species and are eaten by other species. If that relationship of eating and being eaten is so delicate, then why do most ecosystems have so many species? So long as this inconsistency between math and empirical observation exists, we cannot explain one of the most basic features of life on Earth: Why do so many species coexist?

One possibility is that natural selection has created populations of individuals characterized by parameter settings that favor coexistence (Pimentel 1963; Abrams 2000). The other possibility is that the Lotka-Volterra equations are missing some essential phenomena that favor coexistence. Let's follow that second possibility.

SOME EMPIRICAL OBSERVATIONS

In the decades that followed publication of the Lotka-Volterra equations, researchers searched for phenomena that might favor coexistence in a way that the equations had neglected. This search included many experiments on populations of tiny creatures with short lives that could be raised in a lab and observed over short periods of time. I'll describe three such experiments, each yielding worthwhile insight.

Ciliated Protozoans

Just a few years after Volterra's 1926 paper, the Russian ecologist Georgii Gause—while in his early twenties—devised a clever way to evaluate Volterra's equations. He enlisted two species of single-celled creatures known as ciliated protozoans.[5] Entire populations of these creatures can live in a test tube or petri dish of water. Ciliated protozoans are also easy to come by—they're found in just about any scoop of pond water. Many species of ciliated protozoans live by feeding on bacteria. Others are predatory and live by preying on other species of ciliated protozoans. Gause selected *Paramecium caudatum*, which feeds on bacteria, to represent the prey in his laboratory experiment. He selected *Didinium nasutum*, which feeds on those paramecia, to represent the predator.

To prepare his investigation, Gause dumped some oatmeal in water and inoculated the mixture with bacteria. Soon after, the water teemed with bacteria. He strained the oatmeal from the water, leaving a bacteria-rich broth. Then he poured less than a teaspoon of the broth into a small vessel. Into the broth, he dropped 10 paramecia.

If bacteria are plentiful, one paramecium can consume 100 bacteria in an hour. When a young paramecium eats as much as it likes, then in just five or six hours, about half of the paramecium's inside parts migrate to each end of its unicellular body, the cell membrane begins to contract in the middle and then pinches off. In a process that takes just a few minutes, where there had been one paramecium, we are now staring at two half-sized paramecia. Each scoot off in their own direction in search of more bacteria. In 24 hours, the population of 10 paramecia increased to about 30. In 48 hours, Gause's vessel was inhabited by more than 100 paramecia.

At that point, Gause dropped three didinia into the tank. After 24 hours, those 3 didinia had consumed scores of paramecia and increased their own kind to about 25. The abundance of paramecia plummeted from more than 100 to about 35. A day later, the number of didinia held steady, and the paramecium population struggled against extinction. A day later, the paramecia were extinct. The didinia, now with nothing to eat, went extinct the following day. All that remained were bacteria feeding on didinia carcasses and waste products—the recycled parts of bacteria.

Gause failed to reproduce the perpetual cycles that Volterra's equations could predict. Maybe, he thought, the failure arose from an imbalance among Volterra's four fundamental rates (r, a, e, m), canted by some underappreciated simplicity of the microcosm he had created. Maybe, for example, the attack rate, a, was too high and simply outpaced the growth rate of the paramecia, r.

Gause tried again, but this time he aimed to level the playing field. Rather than strain all the oatmeal from the bacteria-laden broth, he left some oatmeal crumbs to sink to the bottom of the test tube. These crumbs, he imagined, would offer a place for paramecia to hide, thereby lowering the attack rate (a).

Into the oatmeal-littered vessel, Gause released about a half dozen individuals of each of the two protozoan species. The number of didinia and paramecia remained about the same for a day. After another day, the number of didinia and paramecia both increased, but not greatly. On day three, the paramecia soared to more than 20, and the didinia began to decline. By day six the didinia had fallen to extinction, and the paramecia had increased to more than 40 individuals. Gause poured the broth down the drain and terminated the experiment.

Undaunted, he reran the experiment. This time Gause dispensed with the oatmeal chunks, but altered the microcosm in another way. On every third day of the experiment, he added three paramecia and three didinia. He did so no matter who seemed to have the upper hand. With this tweak, both populations persisted more than 16 days, and they exhibited cycles reminiscent of those predicted by the Lotka-Volterra equations (Fig. 9.4, panel A). Persisting for 16 days might not seem like much, except that quite a few generations of protozoans passed during that time. To observe a similar number of generations for a vertebrate such as deer would require about two centuries.

Mites

Carl Huffaker ran an experiment during the 1950s that—because of his expertise with agriculture-eating invertebrates—involved oranges, six-spotted mites that feed on the oranges, and predatory mites that feed on the six-spotted mites.[6] Like Gause, Huffaker knew the value of keeping it simple in the beginning. He set 20 six-spotted mites loose in a pile of oranges, and then he set loose two gravid female predatory mites. That's pretty simple. Soon after the release, the six-spotted mite was driven to extinction by the predatory mite, which fell to extinction shortly afterward. The outcome was a rerun of Gause's first take with protozoans.

Huffaker wanted to know whether coexistence might be supported by an environment that was more complicated than just a pile of oranges. So, he designed a more complicated environment. He arranged the oranges on a tray, each orange neatly spaced from its neighbors by a few centimeters, like Christmas ornaments in a box. With a generous application of imagination, Huffaker also dipped the oranges in wax, leaving exposed only a small portion of orange skin from which the six-spotted mites could feed. This was a nice flourish but—to Huffaker's mind—still not sufficiently elaborate. Huffaker erected wooden dowels, rising high above the oranges, to serve as launching points for six-spotted mites searching for the next orange. Still too simple. Knowing that the predatory mites crawl and cannot hop like the six-spotted mites, Huffaker smeared Vaseline on portions of the trays, creating a maze to impede the rate at which predatory mites could search for their next prey. In this fanciful environment, Huffaker had the mites play their deadly game of hide and seek.

A

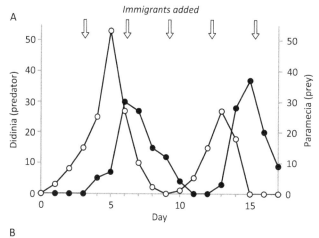

B

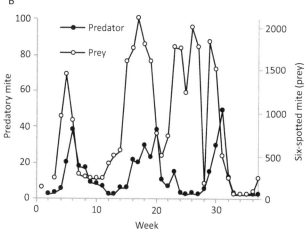

C

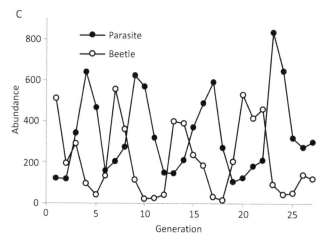

Figure 9.4 Fluctuations in abundance of predators and prey raised in microcosms. The data are from Gause's (1934) experiment with protozoans (A), Huffaker's (1958) experiment with mites (B), and Utida's (1957) experiment with beetles (C). Panels B and C are excerpted from the full duration of those experiments.

Any six-spotted mite that fed from a patch of exposed orange skin that happened to be at that moment occupied by a predatory mite could become prey if it lingered there. But the six-spotted mites were more mobile and quickly moved from orange to orange. Enough of the six-spotted mites were sufficiently swift to evade the predatory mites, but not so many as to completely deprive the predatory mites of a meal. Where the undifferentiated pile of oranges led to quick extinction, this labyrinth of a microcosm supported coexistence. For more than a year, the two populations coexisted, exhibiting cycles not unlike those predicted by the Lotka-Volterra equations (Fig. 9.4, panel B).

With that result, Huffaker demonstrated that persistence does not necessarily require the regular introduction of immigrants. Huffaker's creativity is celebrated as a quintessential example of coexistence depending on details of the physical environment.

Bean Beetles

At about the same time that Huffaker was conducting his experiment, Syunro Utida had been observing predatory relationships involving a beetle that feeds on adzuki beans.[7] This beetle, rather straightforwardly named the adzuki bean beetle (*Callosobruchus chinensis*), is pursued in life by a parasitic wasp that lays its egg on the beetle. Shortly thereafter, a larval wasp hatches and burrows into the beetle, paralyzing the beetle and eating it alive from the inside out. The beetles have three life stages (larval, pupal, and adult). The wasps are effective at attacking beetles only in their larval stage.

Utida's beetles and wasps coexisted in his microcosm for more than 100 generations, exhibiting oscillating cycles throughout (Fig. 9.4, panel C). He offered reason to suppose that their coexistence was favored by the predator's ability to feed on only one of the prey's three life stages.

Recap

The Lotka-Volterra equations led to insight and a problem. The insight is an essential truth that trophic relationships* lead to cyclical fluctuations in abundance. The problem was that the simple mechanisms underlying the Lotka-Volterra equations also indicated that tightly bound trophic relationships were not particularly stable, that extinction was likely, and that coexistence was unlikely. Yet, trophic relationships and coexistence are widespread in nature.

---

* Trophic relationship is a generic phrase covering any relationship that involves one species consuming another, such as predation, herbivory, or parasitism.

The basic solution to that problem was quickly appreciated—namely, nature must possess some phenomena that favor coexistence, but are not represented by the Lotka-Volterra equations. Each of the three microcosm experiments that we considered offers clues about these coexistence-favoring phenomena. With Gause's protozoans it was occasional immigration. With Huffaker's mites it was an environment with refugia where prey can hide. And with Utida's beetles, the condition allowing for coexistence seems to have been selective predation that focused on one age class of prey and spared the other age classes.

In the Wild

Each of the ideas in the previous paragraph has also been observed in the wild. For example, there is a population of tawny owls in northern England that feeds on voles and is preyed on by goshawks. When voles are locally plentiful, more owls tend to arrive from elsewhere. That immigration likely dampens the impact of predation by goshawks (Hoy et al. 2015). Many vertebrate predators (such as wolves) tend to avoid prime-aged prey and prefer to kill calves and senescent prey (Hoy et al. 2021). Refugia are likely important for the dynamics of many prey, including kites preyed on by owls in the Italian Alps (Sergio, Marchesi, and Pedrini 2003), seals preyed on by sharks in the waters off South Africa (Wcisel et al. 2015), and invasive gray squirrels preyed on by pine martens in the United Kingdom (Twining, Montgomery, and Tosh 2021).

The three microcosm experiments described in the previous section have also inspired comparison with two cases, in particular, of animal populations living in larger natures (Fig. 9.5):

· lynx and snowshoe hares of Canada, which exhibit a ten-year cycle.
· wolves and moose of Isle Royale, which exhibit oscillating fluctuations that are at least reminiscent of cycles.

Both cases are comprised of populations that are tightly bound to one another, which is important because that represents the kind of case that Lotka-Volterra equations aim to explain.

These cases contrast with many communities in which multiple species of predators depend on multiple species of prey. In those cases, no one species tends to be so deeply dependent on another. The Lotka-Volterra equations were not intended to describe those more complex cases—at least not without modification.

Furthermore, it is known that cycles occur in more northerly populations of hare, where lynx is not only the dominate predator but also a spe-

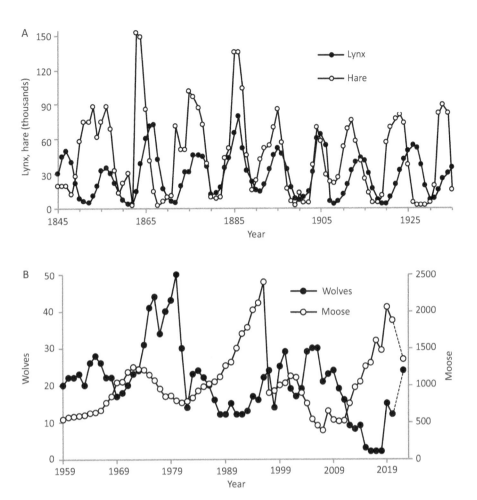

Figure 9.5 Fluctuations in abundances of lynx and hare from Canada and wolves and moose on Isle Royale. Adapted from Odum and Barrett (1971) and Hoy, Peterson, and Vucetich (2022).

cialized predator of hare. But southerly populations of hare tend not to cycle, and they tend to be preyed on by a more diverse set of generalist predators (such as bobcats, foxes, and owls). A similar trend has been observed with voles in Fennoscandia and northern Europe, where cycles tend to occur in northerly populations exposed primarily to a single species of specialist predator, but not in southerly populations where there is a greater diversity of generalist predators (Turchin and Hanski 1997).

MATH AND MECHANISM

Another line of research that followed publication of the Lotka-Volterra equations focused on *mechanisms* within those equations that might be

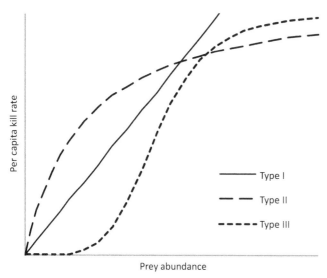

Figure 9.6 Prey-dependent functional responses. These are relationships between prey abundance and per capita kill rate. See main text for details on the different types (shapes) of functional response.

Legend:
— Type I
— — Type II
----- Type III

Axis labels: Per capita kill rate (y-axis); Prey abundance (x-axis)

important for understanding interlinked fluctuations of predator and prey. This line of research has been a mixture of math and empirical observation.

Per Capita Kill Rate

A mechanism that has received much attention is the *per capita kill rate.* The "per capita" part of that label refers to the number prey *each predator* is expected to kill during some given period of time. As such, the units for this mechanism are kills per predator per unit time. This mechanism is sometimes referred to by the shorter expression *kill rate.*

The kill rate is represented in both equations of Figure 9.2 by the product, aN. Kill rate is important because it is the mechanism that links the equations for predator and prey. That is, kill rate is essential for describing gains to the predator population that occur due to losses from the prey population.

The term aN is only one possible expression that can be used to describe kill rate. Volterra used aN because it is simple and because it describes an intuitive and important truth. That is, kill rate should increase as prey abundance increases because prey are easier to find when abundant. This relationship has become known as the Type I *prey-dependent functional response* (Fig. 9.6).[8]

The term aN can be replaced with more complicated expressions if there is reason to do so. One reason is that predators become satiated (full) when they eat enough. Consequently, kill rate should rise with prey abundance (N), but at some point, predators become satiated and further increases in N would lead to slower increases in kill rate. This more complicated response to increasing N is known as a Type II prey-dependent functional response (Fig. 9.6) and represented by this function:

$$KR = \alpha N/(\beta + N), \hspace{2cm} (9.6)$$

where KR is an abbreviation for per capita kill rate. The parameter, α, describes the asymptotic (maximum) kill rate. β describes how quickly kill rate increases toward that asymptote.[9] The right side of Equation 9.6 can replace the two instances of aN in the Lotka-Volterra equations.

There is also a Type III prey-dependent functional response, in which kill rate remains low as N increases from very low densities to account for the idea that predators pursue alternative prey when the abundance of the prey represented by the x axis is rare (Fig. 9.6).

Equation 9.6 is interesting not only because it could be a better description of how predation works, but also because it turns out that replacing aN with $\alpha N/(\beta + N)$ in the Lotka-Volterra equations favors coexistence. That is, the replacement leads to coexistence for a wider range of parameter values, and population cycles tend to dampen over time to constant equilibria (i.e., constant abundances for N and P, not unlike the dynamics in Figure 3.12).[10] So, is the Type II functional response a better description of predation in real populations than the Type I shape?

To answer that question, it helps to consider a specific example. Let's take the wolves and moose of Isle Royale. But first, we'll need a blank graph with N on the x axis and per capita kill rate on the y axis. Recalling that the units on per capita kill rate (i.e., kills per predator per unit time) offers a strong clue as to how we'll estimate kill rate. What my colleagues and I have done on Isle Royale for a number of years is to observe wolf packs from a small aircraft on a daily basis throughout a several-week period during the winter. For example, in 2022, Sarah Hoy and Rolf Peterson observed that the 28 wolves that had been in the population killed 25 moose during a 38-day period (Fig. 9.7).[11] That means the kill rate was 0.71 kills/wolf/month. During that same year, they also counted moose, leading to an estimate of 1,346 moose in the population. Now we can plot that point (1346, 0.71) on the graph. Repeat that process for many years and what emerges is panel A of Figure 9.8, which suggests that a Type II functional response is a better description of kill rate than the Type I functional response for wolves and moose on Isle Royale. That result is representative of a wide range of predators (Jeschke, Kopp, and Tollrian 2004).

The functional responses depicted in Figure 9.6 are three of the many functions developed to describe kill rate. An especially important set of functional responses was developed during the 1980s and 1990s to account

Figure 9.7 Map of Isle Royale National Park showing where wolves from that population's two packs had killed moose in January and February of 2022. Observations were made by following the tracks of wolves in the snow from a small aircraft and form the basis for estimating the per capita kill rate.

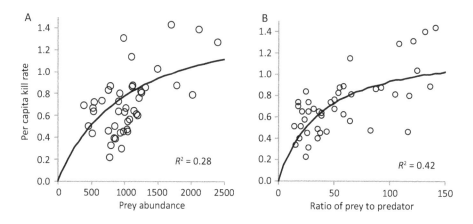

Figure 9.8 Empirical data for the prey-dependent (A) and ratio-dependent (B) functional responses of wolves and moose in Isle Royale National Park, 1971–2022.

for the influence not only of prey abundance, but also of predator abundance (Abrams and Ginzburg 2000). The essential idea is that if N is held constant while predator abundance (P) increases, then per capita kill rate should decline. The reason for this is that predators will spend more time interacting with each other as competition for prey intensifies. This competition among predators could be manifest, for example, as the time it takes to defend a territory from conspecifics. Functional responses that account for the densities of both predator and prey are generally better descriptions of kill rate than those taking account only of prey density (Skalski and Gilliam 2001).

Perhaps the most mathematically parsimonious expression of this idea is the Type II *ratio-dependent functional response*, which calculates kill rate as a function of the ratio of prey to predator (N/P):

$$KR = \alpha(N/P)/(\beta + (N/P)). \qquad (9.7)$$

This is the same as Equation 9.6, except that N has been replaced with N/P.

For the wolves and moose of Isle Royale, the ratio-dependent functional response is a better description of kill rate than the prey-dependent functional response (Fig. 9.8; Vucetich, Peterson, and Schaefer 2002). Mathematical analyses also indicate that predator-dependent functional responses often favor stability and coexistence between predator and prey (Arditi et al. 2004).

Density-Dependent Prey Growth

Another mechanism within the Lotka-Volterra equations is represented by the term rN and the concomitant assumption that prey populations grow exponentially in the absence of predation. It is straightforward to replace rN with the more complex expression $r_{max}N(1 - N/K)$ to represent density-dependent growth (see Eq. 3.2). It turns out that equations with this replacement also predict greater stability between predator and prey.

Numerical Response

The next mechanism to consider is the *numerical response*, which is a loose and flexible term used to described various aspects of how changes in prey abundance (N) or kill rate (KR) lead to changes in predator abundance (P), predator growth rate (dP/dt), or per capita growth rate in the predator population.[12] Thus, Equation 9.5 can be labeled a numerical response.

Observe that Equation 9.5 indicates that per capita predator growth ($dP/dt \times 1/P$) increases linearly with increases in kill rate (aN), where e is the slope and $-m$ is the y intercept of this linear increase. This may seem fine, except that one might suppose that the predator's growth rate is also limited by their physiology and life history. That is, population growth should be limited by the predators being able to live only so long (survival rate) and reproduce only so fast, regardless of the rate at which they acquire food. In other words, predators are limited by a kind of r_{max}. This thinking leads to an asymptotic, rather than linear, numerical response (Ginzburg 1998).

When the right side of Equation 9.5 (which is linear) is replaced with an asymptotic expression, then a wider range of parameter settings lead to coexistence, and cycles tend to dampen over time. Direct empirical evidence for this kind of numerical response in terrestrial vertebrates is sparse because the data are so difficult to gather.[13] One of the few systems where such data are available is the wolves and moose of Isle Royale, where a nonlinear numerical response seems to be a better fit than a linear response (Fig. 9.9). A nonlinear numerical response can also arise, for example, from density-dependent processes, such as territoriality, which has been observed in goshawks preying on tawny owls (Hoy et al. 2017). As with

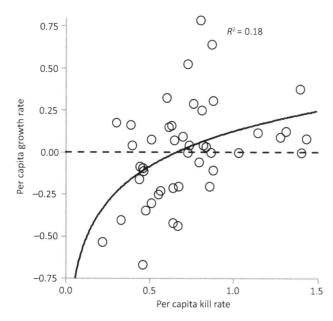

Figure 9.9 Numerical response of wolves in Isle Royale National Park, 1971–2014. While there is a discernable and statistically significant pattern, it is also striking how little variance in growth rate is explained by per capita kill rate. Adapted from Vucetich and Peterson (2004).

several of the other mechanisms that we've been exploring, nonlinearities in the predator population also tend to stabilize predator-prey dynamics.

Recap

Continuing on the heels of the previous recap, we can see that empirical observations support adjusting the mechanisms represented by the Lotka-Volterra equations. These adjustments also tend to favor coexistence by making the predator-prey dynamics more stable, thereby helping to solve the problem described in the previous recap.

PREDATION RATE

Early in the growth of predation knowledge, the Lotka-Volterra equations established kill rate as being of central importance. Consequently, much research focused on answering the questions:

· What causes kill rate to fluctuate over time?
· What are the consequences of those fluctuations in kill rate?

Answers to the first question include various kinds of functional response, and answers to the second question are associated with the numerical response.

Important as the kill rate is, it obscures a subtle, yet important, distinction. The per capita kill rate is an excellent and direct characterization of

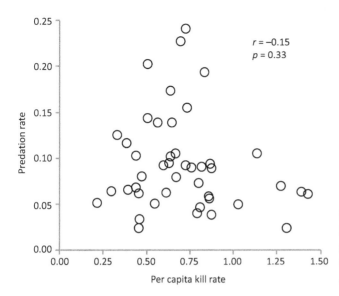

$r = -0.15$
$p = 0.33$

Figure 9.10 The relationship between per capita kill rate and predation rate for the wolves and moose of Isle Royale, 1971–2014. Adapted from Vucetich et al. (2011).

predators' experience, inasmuch as it is the rate at which predators acquire food.[14] A more direct characterization of the prey's experience is the *predation rate*, which is the proportion of the prey population killed by predators per unit time (often a year). In other words, predation rate is a cause-specific mortality rate. While predation rate is a distinct idea, it is not unrelated to the per capita kill rate. In particular, predation rate (*PR*) is equal to the kill rate multiplied by the ratio of predators to prey:

$$PR = KR \times (P/N).$$

With that simple, precise mathematical connection between *KR* and *PR*, one might think that the two quantities contain roughly the same information. In other words, intuition might lead one to think, for example, that a good year for predators (high *KR*) would tend to be a bad year for prey (high *PR*), and vice versa.

Nope. It turns out not to work that way. Mathematical analyses of predation indicate that *KR* and *PR* may be correlated for some systems, but not others (Vucetich et al. 2011). For the wolves and moose of Isle Royale— one the few places where this relationship has been assessed—*KR* and *PR* are essentially unrelated. Some years are good for predators and prey, some years are bad for both populations, and other years are good for one population but not the other (Fig. 9.10).

Given that predation rate represents information that is important to distinguish from the kill rate, we will gain insight by asking two questions about the predation rate. First, how is the per capita growth rate of the prey

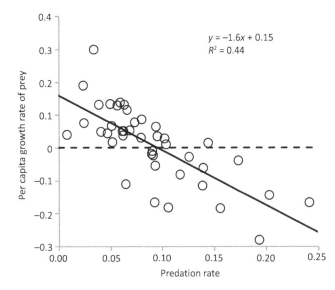

Figure 9.11 Growth rate of the moose population on Isle Royale in relation to predation rate. Notice that growth rate tends to decrease linearly with increases in predation rate, which is usefully thought of as a cause-specific mortality rate. The slope is estimated to be −1.6, which indicates that predation would seem to be more additive than compensatory. The slope may be steeper than additive due to nonconsumptive effects of predation (Zanette and Clinchy 2020).

population (r_t) affected by increases in *PR*? This question is much like asking, as we did in Chapter 8, How does r_t change with increases in the harvest rate (h_t)? Harvest rate and predation rate are similar in that both are cause-specific mortality rates. Furthermore, predation can be additive or compensatory in the same way and for the same reasons that harvest can be. If the slope of the relationship between *PR* (*x* axis) and r_t (*y* axis) is −1, then predation is perfectly additive (Fig. 9.11). If the slope is zero, the predation is perfectly compensatory. The meaning and ecological mechanisms for additive and compensatory predation are largely the same as we discussed for harvest (see Fig. 8.15).

When predation tends toward being additive (slope = −1), then we say that predation exerts a strong *top-down* force on prey populations. When predation tends to be compensatory (slope = 0), then we say the system is more influenced by *bottom-up* processes, in the sense that lower trophic levels are supplying food to higher trophic levels without the lower trophic levels being impacted. If predation is compensatory, then the populations are not tightly linked, and the Lotka-Volterra equations do not apply.

A second question to ask about predation rate is, How does it change with changes in prey abundance (*N*)? The most general answer is that predation rate rises as *N* increases from low values and then deceases as *N* increases further (Fig. 9.12, panel A). Mathematical analyses indicate how details of the functional and numerical response greatly affect the sign (+ or −) and steepness of the slope at various points along the *x* axis. As a result and for example, some populations may effectively spend most of their time on the left side of panel A in Figure 9.12, where the slope is

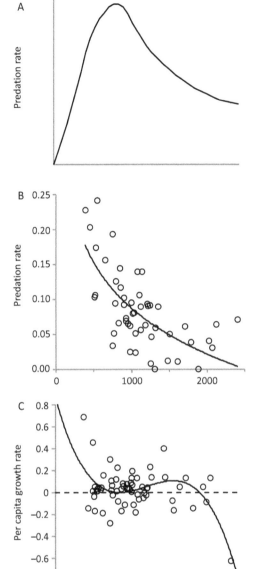

Figure 9.12 Mathematical analyses indicating how predation affects the dynamics of a prey population, illustrated with data from the wolves and moose of Isle Royale. Panel (A) indicates that predation rate can, in principle, rise and then fall as prey abundance increases. For the moose of Isle Royale, predation rate only declines with increasing prey abundance (B), which indicates that predation has a destabilizing influence. That destabilizing influence is very pronounced in the relationship between abundance and per capita growth rate (C). See main text for interpretive details. See online supplement for more details about these data.

positive, and other populations may spend most of their time on the right side where the slope is negative.

Where the slope between N and PR is positive, predation has a stabilizing influence on prey abundance. Conversely, where the slope is negative, predation has a destabilizing influence on prey abundance. To appreciate why, recall that mortality rate (due to all causes) tends to increase with density, which tends to favor negative density dependence. When that is the case the population's dynamics are stable, in the sense that there is an equilibrium and the population tends toward that equilibrium when perturbed from it (Fig. 3.5). However, if an important source of mortality tends to increase as N decreases and if that tendency is strong enough, then the result can be positive density dependence, which you know from Chapter 3 to be an unstable dynamic, in the sense that the population tends to be repelled from, rather than attracted to, an equilibrium.

The question unanswered by panel A in Figure 9.12 is, for any given case, does predation tend to operate on what is effectively the left side of a graph (stabilizing) or the right side (destabilizing)? For the moose of Isle Royale, predation is destabilizing (Fig. 9.12, panel B). That destabilizing influence results in a highly nonlinear density-dependence plot for the moose population (Fig. 9.12, panel C). For a wide range of densities ($\sim500 < N < \sim2,000$) density dependence is positive (unstable). Only at very low densities is r_t very likely to be high (thereby preventing extinction), and only at very high densities is r_t very low (preventing growth to infinity).

When predator and prey are tightly bound, there is a tendency for dynamics to be unstable, as illustrated by the Isle Royale case (see also Fig. 3.10). The tendency for predation to be stabilizing occurs when predator and prey are less tightly bound, as happens when predators have alternative prey and when prey are subject to multiple predators (Gross et al. 2009). We'll see an example of such a case shortly.

MULTIPLE SPECIES OF PREY

The Lotka-Volterra equations, as we've considered them so far, represent the dynamics between a single species of predator and a single species of prey. These dynamics are useful because they pertain, for example, to many predatory insects that are monophagous (that is, foraging specialists that feed on a single species) and some predatory vertebrates that are close enough to being monophagous, such as lynx and hare, stoats and lemmings, and sometimes, wolves and ungulates.

However, most predatory vertebrates are polyphagous. These foraging generalists have diets that are comprised of two or more species of prey. An interesting example of dynamics among one predator and two prey involves eagles (predator), feral pigs (prey), and foxes (prey) on the Channel Islands off the coast of California.[15]

Golden eagles colonized these islands in the 1990s and were primarily supported by an abundant population of feral pigs that humans had introduced in the 1850s.[16] The eagles also preyed *incidentally* on Channel Island foxes, which are an endangered species endemic to those islands.

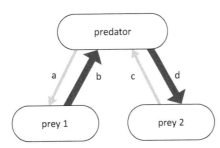

Figure 9.13 Diagram explaining apparent competition. This ecological relationship occurs when the top-down influence of predation is greater for prey 2 than prey 1 ($d > a$) and when prey 1 has a stronger bottom-up influence than prey 2 ($b > c$).

The predation on foxes was incidental only from the eagles' perspective and in the sense that the eagles' diet was largely comprised of piglets (and skunks). While foxes were a relatively unimportant part of the eagles' diet, the impact on foxes was a big deal for the fox population. Shortly after the eagles' arrival, fox abundance plummeted, ending in extinction for two of the six Channel Islands that had been inhabited by foxes.

In other words, a large population of *primary* prey was resilient to predation and subsidized a large predator population. A side effect of that relationship was to adversely affect the *secondary* prey, which was less resilient to predation. In this case, resiliency to predation was due to eagles preying only on piglets (not larger adults),* which are quickly replaced, owing to pigs' ability to reproduce rapidly (several large litters per year). By contrast, the eagles preyed on juvenile *and* adult foxes, who can produce only one small litter per year.

In response to this situation, the National Park Service and Nature Conservancy killed all the pigs on two of the Channel Islands in the 1990s and during the early part of the twenty-first century.[17]

That dynamic among eagles, pigs, and foxes is an example of *apparent competition*. That label refers to the *primary* prey species "appearing" to outcompete the *secondary* prey species. More accurately, the relationship is an indirect consequence of predation that involves two kinds of asymmetry (Fig. 9.13). The first asymmetry is that foxes are more affected by predation than pigs. The second asymmetry is that the eagles benefit more from predation on pigs than on foxes.

A second example of apparent competition is associated with humans having converted large swaths of mature forests in Alberta and British Columbia (Canada) to early seral stages, importantly for the purpose of gas and oil exploration. That habitat conversion led to an increase in the abundance of moose, which led to an increase in the abundance of wolves. The large number of wolves led to high rates of predation on the less abundant caribou (who did not benefit from the habitat conversion).[18] The provinces

---

* A circumstance comparable to Utida's bean beetles.

of Alberta and British Columbia have been responding to this situation by a large-scale, long-term government-funded program to kill wolves. (This relationship between wolves and caribou is related to the discussion about conservation triage and caribou from Chapter 6.)

Other examples of apparent competition include the following:

- Roan antelope in Kruger National Park declined from ~450 to ~45 during the late 1980s and early 1990s. The cause of the decline was elevated lion predation, which had increased because the lions were supported by abundant populations of zebra and wildebeest, which had become abundant in response to artificial watering holes that humans had been maintaining (Harrington et al. 1999).
- Trout have been introduced to many lakes of the western United States. In some of these lakes in northern California, garter snakes prey on the juvenile trout as well as on cascade frogs. Wherever all three species are present, the frogs are rare. But in lakes where trout were never introduced, the frogs are common. It seems that the trout subsidize larger populations of snakes to the detriment of the frogs (Pope et al. 2008).

In each example, the dynamics were triggered by some human influence. Human influence is not a requirement for apparent competition, but it is a trigger for many concerning cases of apparent competition.

Apparent competition is not the only kind of dynamic that can arise among one predator and two prey. The general way to understand such dynamics begins by modifying the Lotka-Volterra equations. The main modification is to develop three equations, one for each species. The predator equation will have a term representing each prey species (in addition to any terms representing density dependence or intrinsic rates of mortality). The other modification often needed for such cases is to use a Type III functional response (Fig. 9.6). This functional response aims to account for the supposition that predators will reduce their effort in pursuing a prey species when it becomes rare in favor of focusing on the more common alternative prey species. We'll see how that kind of functional response affects population dynamics in our next example, which also takes account of multiple predator species.

MULTIPLE PREDATORS (AND MULTIPLE PREY)

Recall that the Lotka-Volterra equations describe a dynamic in which the predator and prey are tightly bound. Just as prey growth is about to explode, predation becomes strong enough to pull the prey back. Then predation seems poised to drive the prey to extinction, except that predator abundance collapses just before prey extinction. If every predator-prey interaction were that intense, it would be very unlikely that prey could endure multiple species of predator. It's not that prey cannot handle high rates of predation—often they can. Rather, it's that the multiple intense relationships would be too delicate, too unstable, and too prone to imbalances leading to extinction. And if the prey were to go extinct first, then the whole system would collapse.

In other words, Lotka-Volterra equations predict that coexistence is increasingly difficult as more species become tightly bound through predation. Yet, the world is full of ecological communities in which multiple species of predator feed on one species of prey. How does that work? What adjustments to the Lotka-Volterra equations are required to more readily result in coexistence? Our next example points the way.

Northern Greenland

Collared lemmings in northern Greenland are preyed on by snowy owls, arctic foxes, long-tailed skuas (marine bird), and stoats.[19] The lemmings also exhibit four-year cycles. Researchers wanted to see if they could explain those cycles by building a model based on their observations from the field.

The centerpiece of their observations was estimates of the functional and numerical response for each of the four predator species. The researchers made their observations during each of 14 years on an area of tundra about $75km^2$ in size. Their observations included counts of animals and nesting sites for lemming, dietary analysis of predators' scat, and direct observations of predatory events. These observations were greatly facilitated by the tundra's open landscape.

The researchers also made observations about the lemmings' relationship with their food supply (vegetation, such as willows and sedges). In particular, they observed that browse rates* were always low even when lemming abundance was high. The tundra habitat also included far more suitable nesting sites than needed to support the lemmings. On the basis of those observations, the researchers supposed that the first term of the prey equation could be rN, representing exponential growth throughout the range of observed lemming densities.

The researchers also observed that three of the four predator species had significant sources of alternative prey on which they relied when lemming

---

* Browse rate is analogous to predation rate. It is the proportion of vegetation (available to be browsed) that was browsed.

density was low. In particular, snowy owls could switch to various species of bird (ptarmigan, waders, and passerines); long-tailed skuas could switch to small birds, fishes, invertebrates, and berries; and arctic foxes could switch to arctic hare, muskox, carrion, birds, eggs, invertebrates, and berries. The ability of those three predators to switch sources of food meant that

· these predators could be relatively abundant even when lemming density was low.
· the functional responses between lemmings and each of these predator species had a Type III shape (Fig. 9.6).
· predation rate by these predators was lowest when lemmings were least abundant, which is opposite the relationship shown in Figure 9.12, panel B, and tends to make predation by these species a stabilizing influence.

All three ideas apply because these predators eat alternative prey when the lemmings are rare. We might say that these predatory relationships are less tightly bound and consequently less delicate and more stable. The last bullet point is also a mathematical consequence of the second bullet point.[20]

The predatory relationship between lemmings and stoats is different. Stoats are specialists in foraging for lemmings and have far less ability to switch prey when lemmings are scarce. The lemming and stoat populations are thus more tightly bound to each other. Furthermore, empirical assessment of the stoats' numerical response indicates a time lag. That is, growth rate for the stoat population at time t depends on their kill rate about six months earlier. When stoats eat more, it takes about six months for that increase to be fully reflected in increased rates of survival and reproduction. You already know that time lags can result in cycles (Eq. 3.5 and Fig. 3.12). Finally, stoats' predation rate tends to decline with increasing lemming density (as in Figure 9.12, panel B), which creates a destabilizing influence on the lemming population.

These researchers went one step further by building equations like the Lotka-Volterra equations, except their set of equations included an equation for each of the five species and the functional and numerical responses matching what they observed in the field for each species. That model produced four-year cycles that closely matched the four-year cycles observed in the lemming population. In that regard, the researchers' equations and the thinking that supports those equations are a highly plausible explanation for why that lemming population fluctuates as it does.

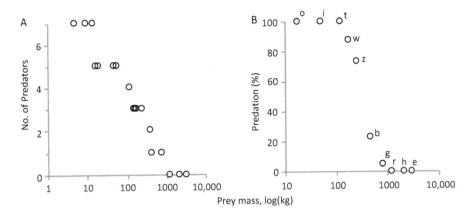

Figure 9.14 Mammalian prey and predators of the African savannas. Mammalian prey of the African savannas are exposed to predation from an increasingly diverse community of predators as body size of the prey decreases (A). That decline is concomitant with an increase in the intensity of predation on the population dynamics with decreasing body size of the prey (B). The y axis of B is the percentage of total mortality that is attributable to predation. The letters in B and the species they represent are o, oribi; i, impala; t, topi; w, wildebeest; z, zebra; b, African buffalo; g, giraffe; r, black rhino; h, hippo; e, African elephant. Adapted from Sinclair, Mduma, and Brashares (2003).

African Savannas

While the ecology of Greenland's arctic tundra is complex, the savannas of Africa are in all likelihood more complex. They are inhabited by about 10 species of predator, ranging in size from wild cats (5 kg) and golden jackals (8 kg) to hyenas (60 kg) and lions (150 kg). These savanna ecosystems are also inhabited by about 17 species of mammalian herbivores, ranging in size from dik-diks (5 kg) and steenboks (10 kg) to hippopotami (2000 kg) and elephants (3000 kg).

Body size seems to be important in mediating predatory coexistence among these species. The smallest species of prey are eaten by as many as seven species of predator. The diversity of predators feeding on any particular prey species declines with prey body size, right on up to the rhinos, hippos, and elephants, who are large enough to be essentially free of predation (Fig. 9.14, panel A). Furthermore, predation is the most important source of mortality for the smallest species of prey and least important for the larger species of prey (Fig. 9.14, panel B).

Those patterns mean that the smallest prey are likely governed primarily by predation dynamics and that the largest prey are likely governed primarily by dynamics involving herbivory. The large number of species interactions involving smaller species of prey on the African savannas also means that there is ample opportunity for their predators to seek alternative prey when any one prey species becomes rare. This prey switching is likely to make the connections involving smaller prey less intense, thereby favoring coexistence (as it did for the lemmings). Among larger species of prey (exposed to a less diverse guild of predators), there is less opportunity for prey switching. In that case, one might expect the dynamics of those prey to be either less stable or stabilized by some other mechanism (perhaps involving their food supply).

CONSUMER-RESOURCE DYNAMICS

While our focus has been predation, the logic represented by the Lotka-Volterra equations can be generalized to represent other trophic relationships, including herbivory, parasitism, and even pathogenic relationships. Doing so typically involves modifying the equations, but the essential logic is the same. This generalized approach to studying trophic relationships is associated with *consumer-resource dynamics* and is of broad importance because most animals consume other organisms, are consumed by other organisms, or both.

Food Chains

That concept of consumer-resource dynamics allows one to consider, for example, the population dynamics of a food chain, comprised of, let's say, a predator, an herbivore, and vegetation. These dynamics can be described with three interlinked equations, one for each trophic level, whose key features include

- equations for the top two trophic levels being linked by the per capita kill rate, as in Figure 9.2.
- equations for the bottom two trophic levels being linked by the per capita browse rate.
- an equation for the middle trophic level, which is influenced predation and herbivory.[21]

Food chains raise a variety of interesting questions—namely, Are fluctuations in herbivore populations driven more by predation or by herbivory? Does more predation mean fewer prey, which in turn leads to more vegetation? Or does more vegetation support more prey, which in turn supports more predators? Can both kinds of dynamics be true at the same time? These questions are associated with the useful, if not vague, over-arching question, To what extent are populations regulated by top-down and bottom-up processes?

Trophic Cascades

As an example of a food chain, let's take the notion that sea otters eat sea urchins and urchins eat kelp. Populations of sea otters are sprinkled up and down the coastal waters of California, British Columbia, and Alaska. Two biologists—James Estes and John Palisano—noticed in the 1970s that where sea otters were living, the urchins were rare, and the rocky reefs were covered in lavish forests of kelp extending 50–100 feet up toward the surface. Where the otters were absent, the forest was replaced with "urchin barrens," where urchins were 10 to 100 times more abundant than they were in kelp forests.

Kelp forests provide architecture that is absent from the denuded urchin barrens. In a kelp forest, the current is slowed, and waves are dampened. Fish, like the rock greenling of the Aleutian Islands, are roughly 10 times more abundant. Glaucous-winged gulls dine almost entirely on fish provided by the otter-tended forest, but when the urchins take over they forage primarily on intertidal invertebrates. Eagles whose nests overlook a kelp forest keep a diet that is roughly an even mix of fish, marine mammals, and sea birds. Eagles living near the urchin barrens forage primarily on sea birds.

Otters compete with, and thereby limit the abundance of, eiders and scoters (diving sea ducks). Otters also eat sea stars, causing them to be less abundant, shorter-lived, and consequently smaller. And the creatures that sea stars eat—mussels and barnacles—therefore experience less predation and live longer, allowing them to grow three to four times faster in otter-maintained kelp forests. Otters are said to be a keystone species because of all the far-reaching consequences of their presence.

You might ask, Why were otters present in some of those places but not others? The answer is that the nineteenth-century maritime fur industry resulted in otter populations being few and far between. Otters are slow to recolonize places from which they have been extirpated, because an otter tends not to travel far from their natal waters. So, what might appear to be an otter-driven dynamic is really a human-driven dynamic.

What I'm describing has been given the label *trophic cascade*. That label sometimes refers to any indirect effect of predation. Other times it refers to far more specific phenomena—namely, when some exogenous force leads to *increased* predator abundance, which results in *decreased* prey abundance, which in turn causes an *increased* abundance of vegetation. A trophic cascade is also said to occur when the previous sentence is re-written by replacing "increased" with "decreased" and vice versa. Because vegetation is such a basic element of any ecological community, appreciable changes in the vegetation can induce changes in the myriad species that use vegetation for food or cover. Finally, trophic cascades are not restricted to predators, herbivores, and vegetation. They can occur among any set of adjacent trophic levels, such herbivorous insects, parasitoids, and hyperparasitoids, or herbivorous fish, predatory fish, and fish that consume predatory fish.

A number of other trophic cascades have been reported. Two impressive ones involved the following:

· Shark populations declined due to overexploitation on the eastern coast of the United States from the 1970s onward. That decline allowed cownose rays to increase by an order of magnitude, which in turn led to a collapse in the abundance of bivalve mollusks.
· Wolves were reintroduced to the Yellowstone ecosystem in the northern Rocky Mountain region of the United States during the mid-1990s, after which there was a dramatic decline in elk of the Northern Range, which sits within the Yellowstone ecosystem. That elk decline was followed by changes in certain vegetative communities and increases in species depending on those communities.

Notice that in each case a trophic cascade was triggered by some process external to the interactions among the three trophic levels. That exogenous influence also happens to have been, for these cases, humans.

Fascinating as these reports are, a more nuanced understanding of trophic cascades is even more fascinating.

Compare, for example, the otter-urchin-kelp cascade with experimental removal of urchins in the more protected waters of the San Juan Channel (in the US state of Washington). Those removals did not result in the increased growth of kelp or other species of macroalgae. Similarly, the kelp forest in the more protected waters of Prince William Sound (Alaska) did not change appreciably when otter abundance declined drastically due to the 1989 oil spill of the Exxon Valdez. These "counterexamples" represent a simple, reliable lesson: not every perturbation of a food chain will result in a dramatic trophic cascade. In the case of kelp, it may be that trophic cascades occur in outer coastal waters, but are less likely to occur in more protected waters.[22]

The shark-ray-bivalve cascade has also been scrutinized. Much of the concern stems from the difficulty in reliably estimating the abundance of marine vertebrates. While some data show that shark abundance declined, those data may represent only a small portion of coastal waters in the eastern United States, and other data that may better represent a larger portion of these waters indicate that sharks declined *and then increased* during the time when rays were reported to have increased. Furthermore, some researchers have argued that:

- Available knowledge about the life history and vital rates of rays indicates that it's not possible for ray abundance to increase as fast as survey data indicate.
- The shark-ray-bivalve cascade does not account for bivalves having been in decline for decades prior to the shark decline, and it's at least plausible that the bivalve decline was largely driven by overexploitation and habitat degradation.
- Shark diet is not sufficiently dependent on rays; nor is shark predation a sufficiently important source of mortality for rays to account for the reported trophic cascade.

As best I can figure, sharks, rays, and bivalves may or may not be involved with a trophic cascade. The scientific process is still sorting it out. But the reliable lesson to draw from this critique is that everything that looks, prima facie, like a trophic cascade, is not necessarily a trophic cascade. An adequate evaluation of a putative trophic cascade requires more evidence than a set of correlated trends in abundance.[23]

Additional scrutiny also leads to a richer understanding of the wolf-elk-vegetation cascade of Yellowstone's Northern Range. As mentioned, elk declined during the decade that followed wolf introduction (1996–2006). But wolves were too rare and elk too abundant during that time for the predation rate to be anywhere close to high enough to have had a significant effect on elk abundance.[24] Furthermore, elk abundance likely declined during that period in response to drought and the intensification of hunting, which coincided with wolf reintroduction. So, if those dynamics were a trophic cascade, then humans, not wolves, were the "predator" that was directly responsible for the decline in elk. By about 2007 and onward, wolves had become abundant enough and elk rare enough for the predation rate to be sufficiently high to keep elk at lower abundances. Aside from humans and wolves, the influence of mountain lions and grizzly bears on elk remains poorly understood. Those predators were part of the predation dynamics before and after the reintroduction of wolves because those predators had not been extirpated from the region.

There has also been an ongoing debate in the scientific literature about the extent to which various vegetative communities on the Northern Range (especially aspen) began to grow in response to the decline in elk.

The concerns include the prospect that other influences, especially precipitation, may have been playing a role in the growth of some of these communities.[25]

Perspective. The most celebrated reports of trophic cascades seem associated with important caveats. I purposefully focused on those caveats, but not to diminish the importance of trophic cascades. Rather, a value in focusing on the caveats is that they represent genuine knowledge about the complexities of trophic relationships.

For example, concluding that macroalgae communities were not appreciably affected by otter removal in Prince William Sound or urchin removal in San Juan Channel does not diminish the finding that otter-urchin-kelp trophic cascades consistently occur in other kinds of marine environments. To the contrary, this "caveat" merely indicates that macroalgae are influenced by more than just urchin herbivory and that the detectable influence of urchin herbivory depends on details that vary among different kinds of marine environments.

Similarly, concluding that Northern Range elk declined from 1996 to 2006 as a result of hunting and drought is *not* evidence that wolves generally have little impact on prey. To the contrary, there is plenty of evidence from scores of studies demonstrating that wolves are generally an important influence on prey dynamics—just not always or in every regard. The richer knowledge comes from understanding when, how, and why wolves are (and are not) a predominant influence.

An important reason that celebrated reports of trophic cascades have been vulnerable to caveats is that those reports arise largely from nonexperimental observation. With that kind of observation, it is difficult (though not impossible) to exclude all the plausible alternative explanations for cascading changes in abundance. Furthermore, it is virtually impossible to conduct an experiment that is *replicated, randomized,* and *controlled*[26] on land- or sea- scapes large enough or for long enough to assess trophic cascades involving creatures such as otters, sharks, and wolves.

But scores of experiments have been conducted in smaller terrestrial systems. These experiments typically involved the removal of birds, lizards, spiders, or ants that prey on herbivorous insects for anywhere from a couple months to a few years and then observing how the plants responded. In 2000, a team of researchers compiled the results of every such study they could identify and found that trophic cascades are definitely the common response to predator removal (Schmitz, Hambäck, and Beckerman 2000; Fig. 9.15).[27] A later experiment showed that when predatory arthropods

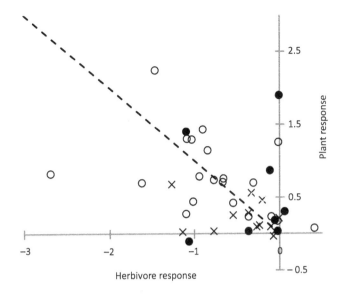

Figure 9.15 Results of predator removal experiments. Each symbol represents the result of a different experiment, with the x and y axes indicating the response of the herbivore and plants, respectively. Smaller values of the x axis indicate a larger decrease in prey abundance, and larger values on the y axis a greater benefit to vegetation. Any experiment appearing in quadrant II of the graph (i.e., above and to the left of the graph's origin) qualifies as a trophic cascade. The different symbols represent different kinds of plant response: tissue damage (○), biomass (x), reproduction (●). Observations below the 45° reference line represent cases in which the effect on plants was attenuated, and observations above the line represent case in which the effect on plants was amplified. Adapted from Schmitz, Hambäck, and Beckerman (2000).

(spiders) were removed from a mesocosm,* the plant community changed so dramatically that restoring the predators did not result in a restoration of the vegetative community (Schmitz 2004). In a word, those mesocosms were not *resilient*, in the sense that we used that term when discussing over-exploitation (Chapter 8).

Recap

The impression seems to be that we have:

· reliable inferences that trophic cascades are a frequent and important phenomenon in small-scale experiments, and
· caveat-riddled inferences about trophic cascades in larger environments.

---

* A mesocosm is typically an outdoor space that is small enough to completely control for the purpose of experimentation.

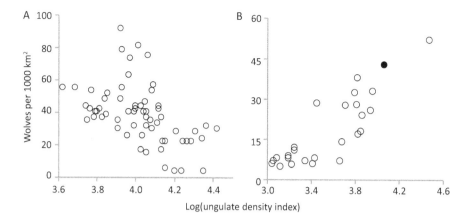

Figure 9.16 Wolf density in relationship to ungulate density for Isle Royale (A) and for 26 sites across North America (B). Each observation in A is a different year. Each observation in B is the time-averaged density for each site. The filled circle in B is Isle Royale. The x axis is an index of ungulate density that allows for comparing sites with different-sized prey (moose, elk, caribou, and so on). The index was calculated following Fuller, Mech, and Cochrane (2003). Adapted from Peterson et al. (2014).

The caveat-riddled inferences result from being unable to perform controlled, randomized, and replicated experiments on large landscapes. Furthermore, human societies seem to care more about trophic cascades on large landscapes than in small-scale environments. It's *almost* as if ecological science is limited to a frustrating trade-off: great answers to impertinent questions or weaker answers to more important questions.[28]

Bottom-Up Processes

While predation (consumption) generally conveys an important top-down influence on community dynamics, it is not always important in that way nor is it the only important trophic influence. Again, wolves and their prey provide an excellent example. On Isle Royale, wolf density is inversely correlated with moose density (Fig. 9.16, panel A), as one would expect if top-down influences were dominant. But for observations taken from sites across North America, the (time-averaged) densities of wolf and their prey were positively associated (Fig. 9.16, panel B), as one would expect if bottom-up influences were dominant. Figure 9.16 is but one way to appreciate that top-down and bottom-up forces are both important—at the same time, but in different ways.

Even at one site, where top-down influences may generally be important, they are not always important. We already spoke of one example, in which the wolf predation in Yellowstone likely had little influence on elk abundance from around 1996 to 2006, but the influence of wolf predation after that time was more important. On Isle Royale, there had been decade-long periods during which predation rates were high enough to significantly reduce moose density and decade-long periods when predation rate was too low to be especially important (Peterson et al. 2014).

RESTORATION OF LARGE PREDATORS

Large terrestrial predators are represented by about 30 species of the order Carnivora—the ones that weigh more than about 15 kg. With respect to conservation, these species are doing very poorly on the whole. More so than other taxa, their populations have been extirpated, their geographic ranges greatly reduced, and their persistence on the planet threatened. Their plight is driven by the difficulty that humans find in coexisting with large carnivores—difficulties such as predators' tendency to kill livestock, kill wild prey that humans hunt, and in some important cases threaten human safety.

Because of those (and other) difficulties, the political will to restore carnivores is limited and tenuous. To encourage political will, advocates develop and advance reasons for why carnivores are important enough to bother making the effort to coexist with them. One of the most frequently offered reasons is that large carnivores are essential to the health and functioning of ecosystems. Evidence provided for that claim often includes oversimplified reports of trophic cascades. Caveats to those reports are taken as undermining the reason to conserve predators. Because these reports are oversimplified and used in politically charged arenas, those opposed to carnivores and those opposed to simplified science tend to raise the caveats.[29] This circumstance is a second reason that the most celebrated reports of trophic cascades are seemingly so vulnerable to caveats.

Insulating the science of trophic cascades from politicization would be greatly aided by developing and offering reasons to conserve carnivores that are not so easily diminished by caveats to oversimplified reports of trophic cascades. For example, a more robust reason to restore large carnivores might arise from an argument built from premises about how humans were wrong to have extirpated carnivores and about humans' obligation to right past wrongs.[30] Such an argument would seem unaffected by any caveat that could possibly be raised about trophic cascades. Recent sociological research also suggests that such an argument is likely more persuasive to the general public than arguments about carnivores' contribution to ecosystem health (Carlson et al. 2023).

In sum, the perspective I'm conveying—for your consideration—involves holding three views, simultaneously: healthy populations of native predators generally have an important influence on ecosystems; some reports of trophic cascades—especially those shared in politically-charged environments—are at least a little oversimplified; and sometimes an ethics-focused argument for some conservation actions is more appropriate than a science-focused argument.

PREDATOR CONTROL

Predator control is a widespread and diverse activity, illustrated by examples such as:

· deterring predators such as racoons and foxes to protect bobwhite nests from predation.
· deploying livestock guardian dogs to protect sheep from wolves.
· killing introduced cats and rats to protect endemic island fauna.
· killing pythons to protect vertebrate fauna in the US Everglades.
· killing wolves, bears, and cougars to assuage some hunters who feel their hunting is upset by those native predators.
· killing seals to protect endangered species of fish and commercial fisheries.
· killing ravens who prey on juveniles of an endangered species of tortoise.
· killing grizzly bears, cougars, tigers, and African lions that have killed humans.

That list is a tiny snippet from a litany of wildly diverse instances of predator control. Predator control is usefully considered to be the altering of predators' behavior or abundance to mitigate or eliminate some undesirable circumstance. The most common undesirable circumstances include killing endangered species, killing livestock, killing wild prey that humans like to hunt, threatening human safety, and inciting fear or hatred. Predator control is also usefully distinguished by whether its focus is the control of a native or non-native predator and by whether the method is lethal or nonlethal.

Predator control perennially raises two questions. First, does it work? Second, do its benefits outweigh the moral costs of the killing, if lethal means are used? A first pass at answering the first question is yes, there is plenty of evidence that prey routinely increase if enough predators are removed, over a large enough area, and for a long enough time—which sometimes means in perpetuity (Lavers, Wilcox, and Donlan 2010; Salo et al. 2010). One would expect that answer given scientific evidence that trophic cascades are a common response to gross, exogenous alterations in predator density. That answer also raises follow-up questions that are nicely illustrated with an example.

In Game Management Unit 13,[31] the government of Alaska has been promoting the killing of black bears, grizzly bears, and wolves for the past 40 years for the purpose of increasing the number of hunted moose. More than 10,000 predators have been killed in the effort.

If the desired, measurable outcome of that killing is an increase in the number of hunted moose, then the effort very likely has been a failure, because correlational analysis of culling and harvest data indicate that increased predator killing has not resulted in increased numbers of hunted moose (Miller, Person, and Bowyer 2022). Moreover, the predator killing has not been conducted in a sufficiently organized manner to allow for much more than correlational analysis.

It is probable that processes other than predation, such as changes in habitat quality, were having a more important influence on the number of harvested moose. It is also plausible that too few predators were killed to expect an increase in the number of harvested moose. But killing enough predators to assuredly increase the number of hunted moose could be both difficult to achieve and undesirable because these predators are native species, and many would say that they deserve to exist free from persecution. Finally, one might argue that the goal had not been to increase the number of hunted moose. Rather, perhaps success was to be indicated by some other objectively measurable criterion. If so, no one has, after four decades of predator control in GMU 13, said what that criterion might be and shown it to have resulted from the predator killing.

This example from Alaska is one of many. Lennox et al. (2018) reviewed more than a 140 cases of predator control *against native predators*. They concluded that only about a quarter of those cases could be counted as successful, if success is defined as killing enough to meet a stated objective, but not so many as to unduly impair the predator population. The causes of failure were diverse, including:

- inadequate design and monitoring to know if success had occurred.
- failure to reduce predator abundance by enough or for long enough to result in the stated objective.
- killing enough predators to release meso-predators* in a way that prevented realization of the objective.

---

* Meso-predators are typically middled-sized predators, such as foxes and raccoons. While they are predators, they do not occupy the top trophic level in a food chain or food web, as is the case for an apex predator.

- black swan events[32] and other unforeseen consequences, such as when the removal of predators had the indirect effect of increasing disease-related deaths in the prey population (Packer et al. 2003).

The conclusions of Lennox et al. (2018) suggest the appropriateness of replacing the question—does predator control work?—with a more robust set of three questions. Each of these questions is listed below, followed by some considerations for how they might be answered.

1. *Can a reasonable outcome be realized with nonlethal methods?* If yes, then it would seem that lethal methods should not be used. A concern in answering this question is that nonlethal control is often dismissed without adequate consideration. This dismissal is premature, given that lethal control fails so frequently and that nonlethal control may be applicable to a wide range of scenarios (Palmer et al. 2005; McManus et al. 2015; Stone et al. 2017).

2. *If a reasonable outcome cannot be realized with nonlethal control, then is the plan for lethal control likely to appreciably mitigate the problem that the predators are causing?* The frequent failure of lethal control against native predators suggests that the answer to this question will often be no and that when an affirmative answer is attainable, it will require more robust planning than is commonly applied. A basic obstacle for predator control is attenuation of the consequences of the lethal control as they work through the chain of processes that end with the desired outcome—often hunters' satisfaction (Fig. 9.17). The take-away from Figure 9.17 is that socio-ecological systems are not precise machines with dials that can be turned a little (kill some predators) and get a good result (e.g., happy hunters) or turned more (kill more predators) and get a better result (e.g., happier hunters).

   Step 2 in Figure 9.17 is especially worth highlighting. Increased prey abundance results from reduced predator abundance only to the degree that predation is manifesting a top-down influence. While top-down influences are generally important, they are not always important (as we saw in the section on "Bottom-Up Processes" and Figure 9.16). Furthermore, reliable judgments about when top-down influences are relatively strong are typically available only after the dynamics have occurred, by which time the dynamics may or may not have changed.

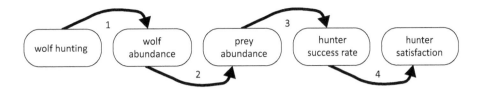

Figure 9.17 Key socio-ecological processes that intervene between killing native predators and hunter satisfaction for a scenario where such satisfaction is the hoped-for outcome of killing predators. Numbered arrows represent processes that connect boxes, which represent elements within the system. With respect to 1, Chapter 8 indicates that harvest strategies do not typically result in precise outcomes with respect to abundance. Furthermore, one should not unduly risk killing so many predators as to impair the health or functioning of predator populations. With respect to 2, this chapter indicates that specified changes in predator abundance do not lead to precisely predictable numbers of prey. With respect to 3, hunter success depends on many more factors than prey abundance (e.g., Cooper et al. 2002). With respect to 4, hunter satisfaction depends on much more than their success rate (Bradshaw et al. 2019).

3. *If nonlethal control is unlikely to work and lethal control is likely to work, will its benefits outweigh the moral costs of the killing?* Because this is a difficult question to answer, it is important to appreciate that it doesn't need to be answered if the first question is answered yes, or the second question is answered no. When the question does need to be addressed, the only robust way to do so is by the methods of argument analysis presented in Chapter 4.

CONCLUSION

Our journey through predation highlighted two questions that have motivated predation research:

· What processes allow populations of predators and prey to coexist when intuition suggests that their coexistence should be tenuous?
· For what reason and by what means should humans learn to coexist with predators?

The first question is one of the most basic in all of ecology, and the second question is one of the thorniest in all of conservation. So, you won't be surprised to know that we've just scratched the surface of both questions. But you can be satisfied to know that this chapter has prepared you to pursue those questions or any idea about predation to which you set your mind.

NOTES

1. Why Study Animal Populations?

1. Kill rate and predation rate are discussed in detail in Chapter 9.
2. The following passages are excerpted and lightly edited for context from *Restoring the Balance* (Vucetich 2021), a book about my experiences researching wolves on Isle Royale.
3. That statistic corresponds (for a typical-sized pack of, say, six wolves) to 4.5 moose per month for the entire pack, or about one moose every seven days. These statistics also pertain to kill rates during the winter months. The best evidence suggests that kill rates of moose (older than about 9 months of age) during the summer are only half. For technical details, see Vucetich et al. (2011).

2. Proportional Growth and Density-Independent Dynamics

1. Many animals, including birds and mammals, are morally relevant to the point of being persons—not humans, but persons. Many well-reputed scholars find it inappropriate to refer to a person as "it." In this book, I'll refer to animals as "they" when the sex of the animal is not known. For the same reason, I refer to elephants in this sentence as "who" rather than "which." For more on the personhood of non-human animals, see Chapter 1 of Vucetich (2021) and Gruen (2021).
2. Estimates of abundance in 1945 and 1946 were 407 elephants and 550 elephants, respectively. Keep in mind, these are *estimates* of abundance. And it's not biologically plausible for an elephant population to increase by 40% in one year. Thus, it's useful to suppose that 407 is an underestimate and 550 is an overestimate.
3. I am using these percentages (2% and 7%) for heuristic value. While they approximate elephant dynamics, more precise information about the demography of elephants can be found in van Aarde, Whyte, and Pimm (1999), Moss (2001), and Wittemyer, Daballen, and Douglas-Hamilton (2013).
4. "Exo-" refers to outside, and "-geneous" is related to "genesis."
5. In case you didn't know, 7% is equal to 0.07, and 2% is equal to 0.02. Those are different ways of writing the same values.

6. For some purposes, this would be an overly simple way to estimate density because the area inhabited by elephants within Kruger NP could change over time. Nevertheless, this calculation of density is adequate for our purpose here.

7. Noting that elephants immigrating into Kruger NP likely had a nonnegligible influence for some periods of this history of growth does not obviate the notion that growth for this period had been essentially density-independent, though it would alter one's impression for rates of mortality and recruitment.

8. For source material on this crane case, see Butler, Harris, and Strobel (2013) and US Fish and Wildlife Service (2018).

9. Strictly speaking, the r in Equation 2.1 is the annual per capita growth rate, and the r in Equation 2.4 is the instantaneous per capita growth rate. While there is an important difference between the two that I highlight with a practice problem in the online supplement, they can be thought of as virtually the same, insofar as they help us understand how populations grow by proportions.

10. For source material on the deep-sea fish example, see Devine, Baker, and Haedrich (2006) and Baker, Devine, and Haedrich (2009).

3. Density-Dependent Dynamics

1. For more on how life history influences density dependence, see Fowler (1981).
2. For source material on this case, see Brashares, Werner, and Sinclair (2010).
3. For source material on this case, see Bourbeau-Lemieux (2011).
4. For source material on the life history of gray-sided voles and bank voles, see Hansson (1985).

4. Ethical Dimensions

1. The strange-sounding name for this kind of logical fallacy refers to formal properties of logical arguments that are beyond the scope of this book, but are easily understood with a little reading that would follow an internet search of the phrase, "fallacy of the undistributed middle."

2. We could start with the opposite conclusion—namely, that elephants should not be culled when they are overabundant. Doing so would lead us to the same insight, but also lead to the use of many grammatical negatives (e.g., not be culled) which tend to make arguments slightly more difficult to follow.

3. The first notion of ecosystem health is invoked when one says, for example, "this forest is healthy because it provides enough timber for humans." The second notion of ecosystem health is invoked when one says, for example, "this forest is not a healthy ecosystem because it has been affected too much by humans." For more see the section, "Ecosystem Health" and Chapter 9 of Vucetich (2021).

4. Evaluating P4 would require drawing on the insights of a branch of ethics known as *deontology*, which indicates that some actions are wrong, no matter the consequences. Evaluating P4′ would require drawing on a branch of ethics known as *consequentialism*, which indicates that the rightness of an action is determined by some weighing of its consequences.

5. While this view of ecosystem health seems to be held by many, it has a critical weakness that rises from depending on the naturalistic fallacy and an untenable dichotomy between what is natural and unnatural. For more, see Chapter 9 of Vucetich (2021).

6. These statements are as importantly true if the phrases "what we value" are replaced with "how we behave."

5. Structured Populations

1. For source material about the orca J2, see Podt (2017).
2. Animals that are zero years old are less than 12 months of age, and animals labeled "four years old" are between their fourth and fifth birthdays.
3. i and j are what mathematicians refer to as *indexing variables*. For our purposes, think of i and j as stand-ins for any pair of specific integers. In other words, the statement to which this note is attached is a *generic* way of saying, for example, that $a_{1,3}$ refers to element of the first row and third column or that $a_{2,4}$ refers to the element of the second row and fourth column.
4. This is the closest I've come to indicating that these models of structured population dynamics are also density-independent models. More on this soon.
5. Except for the (mathematically) trivial case where the population declines, in which case the equilibrium is zero.
6. The online supplement provides a toehold for learning about structured populations that are density-dependent.
7. I realize that I just indicated that these populations do not have an equilibrium *abundance*. But we are about to learn that a population's λ is dynamic. It can change over time, and these changes involve a tendency to approach an equilibrium value for λ.
8. Strictly speaking, it is not quite correct to speak about 5% increases when speaking about elasticities. Rather, elasticities require the language of calculus, which refers to the influence on λ of an infinitesimally small change in some element of A. Regardless, we will use Excel to solve problems that yield valuable insight using arithmetic (not calculus) to make predictions about percentage changes whose magnitude are whatever makes ecological sense, including, for example, a 5% change.
9. For source material to this case, see Sæther and Bakke (2000).
10. For source material about this case of Laysan albatross, see Finkelstein et al. (2010).
11. As of the writing of this note in March 2024, the last sighting of Wisdom to be reported on the internet was December 2023.
12. This sentence can have two meanings. One is a simple statement of fact, and the other is worth reflecting on.
13. North Atlantic right whales are one of three species of right whales. Hereafter, I'll refer to this species by the shorter phrase "right whale."
14. For source material about this case of right whales, see Fujiwara and Caswell (2001) and Kareiva (2001).
15. For source material about this case of polar bears, see Hunter et al. (2010).

6. Extinction Risk

1. Source material for the statistics and claims in this paragraph and the next are from Pimm et al. (2014), Hoffman et al. (2010), and Rodrigues et al. (2014).
2. For a reminder of environmental stochasticity, see Chapter 3. The depressive effect can be illustrated with a simple example. Suppose abundance declines

from N_t to N_{t+1} due to r_t being equal to some value that we'll call, $-X$. Suppose that the next year's r_t is $+X$. If that were the case, N_{t+2} will be less than N_t. In other words, the average value of r_t for this (short) time period is 0 (because the average of $-X$ and $+X$ is 0), but the Var[r] was greater than 0, so the population declined. The online supplement guides you through an exercise to demonstrate this idea. Briefly, it involves setting $N_o = 100$ and $X = 0.13$, projecting the dynamics I just described in Excel over a longer period of time, and observing the population decline. The depressive effect of environmental stochasticity on long-term growth rate is quantified as $r' = E[r] - Var[r]/2$, where r' is the long-term average growth rate (Lande 1993, Caughley 1994).

3. It may seem easy enough to intuit what it means to ask a computer to pick a set of random numbers. In this case, we also want the random numbers to have certain statistical properties. In particular, we want the average of these numbers to be approximately equal to E[r]. Furthermore, we want these numbers to have a certain variability or "spread" around the average value. More specifically, we want a spread like that of a normal (or bell-shaped) distribution. That spread is defined by the variance (or standard deviation, which is the square root of the variance). For a bunch of numbers drawn from a normal distribution, 68% of these numbers will be within one standard deviation of the average, 95% will be within two standard deviations, and 99.7% will be within three standard deviations.

4. There are various ways to ask a computer program (like Excel or Matlab) to constrain N_t in relationship to N_{max}. The online supplement includes an Excel-based tutorial that walks you through one of these ways.

5. Due to the stochastic nature of these models, different runs may lead to slightly different answers.

6. For details, see Haines et al. (2005).

7. For a reminder about r_{max}, see Equation 3.2 and Figure 3.3. Also the details of Fagan et al. (2001) are different from what I've been describing, but similar enough to be presented as count-based PVAs. Furthermore, Fagan et al. did not, strictly speaking, estimate Var[r], but what they did is close enough to be presented as such for pedagogical purposes.

8. This idea is considered with the online supplement for Chapter 3 in an exercise about the marmots of Vancouver Island (see also Fig. 3.9).

9. We'll have a close look at the influence of genetic processes, such as inbreeding depression, on extinction risk and MVP size in Chapter 7.

10. But see the section on "Resiliency to Overexploitation" in Chapter 8.

11. Larger-bodied species tend to be longer-lived.

12. For more insight into how multiple populations distribute risk, see, for example, Sneath (2018).

13. Interference competition is an ecological relationship where an organism directly interferes with another organism's access to resources, such as food. Territoriality and other aggressive behaviors are examples. Interference competition is often contrasted with exploitive competition, where access to resources is limited by the one organism using those resources faster than the other organism.

14. For context, the cost of a fully funded US endangered species program would be $3.67 per American per year, while the cost of the US defense budget is equivalent to $2,286 per American per year. To see how stingy the American

government is with conservation, compared to other nations, see Lindsey et al. (2017).

15. We'll see how oil and gas extraction led to increased wolf predation in the portion of Chapter 9 that explains *apparent competition*.

16. For context, the estimated cost of conserving all of the herds is less than the expected cost of building and maintaining one F-35 fighter jet (Roblin 2021).

17. For more, see Burgman (2004), McCarthy et al. (2004), Krueger et al. (2012), Martin et al. (2012), and Heeren et al. (2017).

18. SAFE is an acronym created by Clements et al. (2011) from the phrase "species' ability to forestall extinction."

19. A more complete truth is that a full description of extinction risk requires placing a mark on the graph for every point in time (x axis). The result is a line known as the cumulative distribution function for times to extinction (Vucetich and Waite 1998). For the sake of simplicity, we'll continue to focus on describing extinction risk with a single point.

20. Even though abundance, by itself, is often a poor indicator of extinction risk (Fig. 6.2) and the values in Table 6.2 seem at odds with analyses aimed at finding a universal MVP.

21. For more on extinction risk and geographic range, see the online supplement. The importance of large geographic range is also implied by the tendency for populations to appear in region C of the graphs in Figure 6.3.

22. For more on various kinds of rarity, see Yu and Dobson (2000).

23. For an annotated list of references pertaining to this section, see the online supplement.

24. A more complete expression of this idea is that the meanings of those words—especially the relationship of those words to one another—are remarkably nuanced, flexible, or both. While this section offers a sense of this complexity, fuller accounts of hope and fatalism may be found in a list of references provided in the online supplement.

7. Genetic Processes

1. In this chapter, we are focusing on the nuclear DNA of sexually reproducing, diploid organisms with two sexes. But, note: animals also have mitochondrial DNA, and sexual reproduction is not the only mode of reproduction among animals.

2. An allele is one of two or more alternative forms of a gene found at a particular locus (or location on a chromosome). See Figure 7.4 for an example.

3. *Phenotypical plasticity* refers to the amount by which a phenotype can change—given its genetic make-up—in response to environmental conditions. For example, body size depends on an individual's genetic make-up, but also their nutritional condition.

4. While one generally expects randomly selected individuals from a larger population to be unrelated, this may not be true if the population had experienced a bottleneck. European bison are an example of this concern. They had been reduced to 12 individuals, but have since increased to about 7,500 individuals.

5. Dominant and recessive represent just two of several patterns of phenotypic inheritance. For a broader survey of inheritance patterns, search that phrase on the internet.

6. It is quite worthwhile to look at Figure 2 in Massaro et al. (2013), which shows the black robin pedigree.
7. Population geneticists often use a telegraphic definition for F—that is, the probability of "inheriting two alleles that are identical by descent"—to stand for the longer expression I used,—namely, "inherit two alleles . . . appeared in an ancestor's genome."
8. More insight on the connection between allele frequencies and heterozygosity comes from the Hardy-Weinberg principle, about which more can be learned with a quick search of the internet.
9. There is also an approximate sense by which the severity of inbreeding depression is represented by the distance from that reference line. Later in the chapter I'll present a better way to quantify the severity of inbreeding depression.
10. I have been writing as though the expression of deleterious recessive alleles is the only mechanism that gives rise to inbreeding depression. In truth, inbreeding depression can also arise from a process known as *heterozygotic advantage*, which refers to a fitness advantage that may occur among individuals with higher levels of heterozygosity. Evidence suggests that this is a less important mechanism for inbreeding depression (Charlesworth and Willis 2009; but see, Bensch et al. 2006).
11. This is a useful simplification of how VORTEX handles inbreeding and inbreeding depression. For details, see Lacy, Miller, and Traylor-Holzer (2021) and O'Grady et al. (2006).
12. The explanation for this result is analogous to a finding from Chapter 6 (Fig. 6.2), where we learned that decreasing N_{max} does not much affect extinction risk if E[r] is negative. Here, the idea is that taking account of inbreeding depression in a population that is declining (for other reasons) doesn't seem to make the population go extinct quicker, because the population is already quickly headed to extinction.
13. Because the idea is so abstract, it is useful to see the same idea expressed differently. Jamieson and Allendorf (2012) write: "Heterozygosity is lost because of genetic drift at a rate of $1/2N$ per generation in an 'ideal' population of N individuals; that is, one of a constant size in which the next generation is produced by drawing $2N$ genes at random from a large gamete pool to which all individuals contribute equally" (579).
14. This equation is a simplification of Equation 3 from Nunney (1996), which assumes that age structure and maturation time are identical for both sexes.
15. While I describe F-statistics in terms of probabilities, there are other equally accurate interpretations. In particular, Allendorf et al. (2022) provide a well-written interpretation of F-statistics in terms of Hardy-Weinberg equilibrium, where F_{IS} quantifies deviation from Hardy-Weinberg proportions within a local population. F_{ST} is a measure of divergence in allelic frequency among populations; and F_{IT} is a measure of overall departure from Hardy-Weinberg proportions in the entire population due to F_{IS} and F_{ST}.
16. A wolf generation is about four years.

8. Exploited Populations

1. For context, the average European consumes a similar amount of meat annually (Food and Agriculture Organization of the United Nations 2020).

2. A declining trend in the number harvested is an indication of a declining trend in abundance if (among other conditions) hunters' efforts are constant or increasing as abundance declines. That condition is thought to hold in this case.
3. If density dependence is nonlinear, then you'd also need an estimate of θ.
4. Environmental stochasticity can be added to any of several elements of a density-dependent population. For example, one could suppose that K fluctuates from year to year. In Equation 8.3, environmental stochasticity is added to the product, $r_{max}(1 - N_t/K)$, which means that environmental stochasticity is adjusting each year's per capita growth rate (r_t).
5. The is an example of the general effect that time delays have on population dynamics (Fig. 3.12).
6. Setting a sensible threshold depends on the population's dynamics, but envision the threshold being set to some value like K or $\frac{3}{4}K$.
7. For more on this topic, search the internet for jargon used in this sentence, i.e., "future discounting" and the two phrases in quotations.
8. Source material for this example is from Milner-Gulland et al. (2001, 2003).
9. While life history trade-offs are common, they are not universal. In some cases, an individual can have both high survival and high reproduction (e.g., Pelletier, Hogg, and Festa-Bianchet 2006).
10. In the end, hunters killed 23 wolves.

9. Predation

1. The precise meaning of a, expressed in English words, would be a little abstract. However, it is adequate to know that a higher value of a means that the predators find prey more quickly due to some behavior or trait of the predator, the prey, or both.
2. There is more than one reasonable interpretation of m. An alternative interpretation begins by recognizing that predator abundance should be adjusted downward to account for predators dying from causes having nothing to do with food limitation, such as accidents, disease, or old age.
3. Alfred Lotka (1925) independently discovered the same equations and published them a year earlier than Vito Volterra (1926). They used different reasoning to arrive at their conclusions. Both are given equal credit for the discovery.
4. The online supplement includes an exercise in which you can create Lotka-Volterra population dynamics in a spreadsheet.
5. This research is described in Gause (1934). Also, while the subject of this book is animal populations, note that protozoans are not animals. Protozoans can locomote, forage, and evade predators. They are too sophisticated to qualify as bacteria, yet too simple (proto-) to qualify as animals (-zoans).
6. The research is described in Huffaker (1958).
7. This research is presented in Utida (1957).
8. The phrase "functional response" is shorthand for a formal way of speaking that indicates that kill rate is the *response* to a mathematical *function* that depends, in this case, on prey abundance (N).
9. More specifically, β is the value of N where KR is half the maximum value. Equation 9.6 is only one way to write an equation with this Type II shape. This shape can also be justified by alternative ecological reasoning (Dawes and Souza 2013).

10. The online supplement includes an exercise featuring this idea.

11. See Hoy, Peterson, and Vucetich (2022).

12. Using the notation of differential equations, the per capita growth rate for predators is $dP/dt \times 1/P$. That somewhat complicated-looking notation is analogous to r_t in the context of difference equations (e.g., see Equation 2.2).

13. By "this kind" I mean the relationship between kill rate and predator growth rate. Other kinds of numerical response, such as the relationship between N and P, are more common.

14. Notwithstanding that kill rate does not account for the energetic cost of acquiring prey (which may vary with ecological conditions) or losses due to scavengers (Vucetich, Peterson, and Waite 2004; Vucetich, Vucetich, and Peterson 2012). In this sense kill rate is a *gross* rate of acquisition, not a *net* rate. Per capita kill rate also represents what is referred to as *scramble competition*, whereby each predator is assumed to get similar amounts of food. Some instances of *contest competition*—whereby some predators get enough food to survive and reproduce, but others do not—may not be well represented by the per capita kill rate (Royama 2012).

15. Source material for this case is Roemer, Donlan, and Courchamp (2002) and Courchamp, Woodroffe, and Roemer (2003).

16. It's not well understood why golden eagles arrived so many years after pigs had become feral. Bald eagles and their territorial behavior may have excluded golden eagles, until bald eagles were extirpated in the 1960s due to hunting and DDT. This possible explanation was given by Latta (2001).

17. For more, see Parkes et al. (2010).

18. You can learn more about apparent competition among wolves, caribou, and moose in McLellan et al. (2010).

19. Source material for this case is from Gilg, Hanski, and Sittler (2003) and Gilg et al. (2006).

20. See online supplement for details.

21. An exercise in the online supplement walks you through the development of a diagram like Figure 9.2 for a food chain. That exercise is the sure way to understand these three bullet points.

22. Source material for otter-urchin-kelp cascades is from Estes, Peterson, and Steneck (2010), Carter, VanBlaricom, and Allen (2007), and Dean et al. (2000).

23. Source material for the shark-ray-bivalve case is from Heithaus et al. (2008) and Grubbs et al. (2016).

24. Recall that $PR = KR(P/N)$. So, if P/N is low (as it was during the first decade after reintroduction), then PR will be low. Annual PR of recruited elk during this period was likely never higher than 0.05/year, and during some of those years it would have been much lower than 0.05.

25. Source material for these ideas include Vucetich, Smith, and Stahler (2005), Ripple et al. (2014), and Peterson et al. (2014).

26. An experiment is *controlled* if the outcome of an experimental treatment is compared to a control, which is treated identically to the experimental treatment in every way except the application of the treatment. An experiment is *replicated* if there are multiple trials of the treatment and control. An experiment is *randomized* if the subjects receiving the treatment (as opposed to being controls) are selected at random. Experiments with those properties produce the most robust inferences about whether the experimental treatment truly

caused the observed results. Some experimental designs have subsets of those features; their ability to infer causal relationships tends to be weaker than experiments with all of those features.

27. Numerous and various kinds of experiments have also been performed in marine, aquatic, and microbial communities. That research gives an impression similar to what I am conveying about terrestrial communities.

28. While that perspective contains a nugget of truth, it is also a wildly inadequate understanding of how science contributes to the growth of knowledge. A more measured perspective is that some scientific questions are harder to answer than others and that the harder questions are often of special interest.

29. The two kinds of opponent are often *not* the same people.

30. The online supplement includes an exercise that involves turning reasons like this into formal arguments, using the principles of Chapter 4, so that they can be evaluated for soundness and validity.

31. About the size of Latvia or West Virginia.

32. Black swan events are extremely important and just as unexpected (Taleb 2007).

REFERENCES

Aanes, Sondre, Steinar Engen, Bernt-Erik Sæther, Tomas Willebrand, and Vidar Marcström. 2002. "Sustainable harvesting strategies of willow ptarmigan in a fluctuating environment." *Ecological Applications* 12, no. 1: 281–290.

Abrams, Peter A. 2000. "The evolution of predator-prey interactions: Theory and evidence." *Annual Review of Ecology and Systematics* 31: 79–105.

Abrams, Peter A., and Lev R. Ginzburg. 2000. "The nature of predation: Prey dependent, ratio dependent or neither?" *Trends in Ecology & Evolution* 15, no. 8: 337–341.

Alaska Department of Fish and Game. 2014. "Subsistence in Alaska: A Year 2014 Update." https://www.adfg.alaska.gov/static/home/subsistence/pdfs/subsistence _update_2014.pdf.

Alexander, Kristina. 2010. "Warranted but precluded: What that means under the Endangered Species Act (ESA)." Washington, DC: Congressional Research Service. https://esadocs.defenders-cci.org/ESAdocs/misc/CRS_Warranted_but _precluded.pdf.

Allen, Summer. 2018. "Eight reason why awe makes your life better." *Greater Good Magazine*. Berkley: The Greater Good Science Center at the University of California, Berkeley. https://greatergood.berkeley.edu/article/item/eight _reasons_why_awe_makes_your_life_better.

Allendorf, Fred W., W. Chris Funk, Sally N. Aitken, Margaret Byrne, and Gordon H. Luikart. 2022. *Conservation and the Genomics of Populations*. Oxford: Oxford University Press.

Allendorf, Fred W., Gordon H. Luikart, and Sally N. Aitken. 2012. *Conservation and the Genetics of Populations*, 2nd ed. Chichester, UK: Wiley-Blackwell.

Altmann, Jeanne, and Susan C. Alberts. 2003. "Intraspecific variability in fertility and offspring survival in a nonhuman primate: Behavioral control of ecological and social sources." In *Offspring: Human Fertility Behavior in Biodemographic Perspective*, edited by Kenneth W. Wachter and Rodolfo A. Bulatao, 140–169. Washington, DC: National Academies Press. http://www.nap.edu/catalog/10654 .html.

Arditi, Roger, Jean-Marc Callois, Yuri Tyutyunov, and Christian Jost. 2004. "Does mutual interference always stabilize predator–prey dynamics? A comparison of models." *Comptes Rendus Biologies* 327, no. 11: 1037–1057.

Ariño, Arturo, and Stuart L. Pimm. 1995. "On the nature of population extremes." *Evolutionary Ecology* 9, no. 4: 429–443.

Arponen, Anni. 2012. "Prioritizing species for conservation planning." *Biodiversity and Conservation* 21, no. 4: 875–893.

Baker, Krista D., Jennifer A. Devine, and Richard L. Haedrich. 2009 "Deep-sea fishes in Canada's Atlantic: Population declines and predicted recovery times." *Environmental Biology of Fishes* 85, no. 1: 79–88.

Bauer, Hans, Craig Packer, Paul Funston, Philipp Henschel, and Kristin Nowell. 2016. *Panthera leo.* The International Union for Conservation of Nature Red List of Threatened Species. Gland: IUCN. https://doi.org/10.2305/IUCN.UK .2016-3.RLTS.T15951A107265605.en.

BBC (British Broadcasting Company). 2020. "Botswana auctions off permits to hunt elephants." February 7. https://www.bbc.com/news/world-africa-51413420.

Beissinger, Steven R. 1995. "Modeling extinction in periodic environments: Everglades water levels and snail kite population viability." *Ecological Applications* 5, no. 3: 618–631.

Bell, Donovan A., Zachary L. Robinson, W. Chris Funk, Sarah W. Fitzpatrick, Fred W. Allendorf, David A. Tallmon, and Andrew R. Whiteley. 2019 "The exciting potential and remaining uncertainties of genetic rescue." *Trends in Ecology & Evolution* 34, no. 12: 1070–1079.

Bensch, Staffan, Henrik Andrén, Bengt Hansson, Hans Chr. Pedersen, Håkan Sand, Douglas Sejberg, Petter Wabakken, et al. 2006. "Selection for heterozygosity gives hope to a wild population of inbred wolves." *PLoS ONE* 1, no. 1: e72.

Bergerud, Arthur T. 1974. "Rutting behaviour of the Newfoundland caribou." In *The Behaviour of Ungulates and Its Relation to Management*, edited by V. Geist and F. Walther, 395–435. Morges, Switzerland: World Conservation Union.

Bijlsma, R., J. Bundgaard, and A. C. Boerema. 2000. "Does inbreeding affect the extinction risk of small populations? Predictions from *Drosophila*." *Journal of Evolutionary Biology* 13, no. 3: 502–514.

Blascovich, Gregory Major, and Alexander L. Metcalf. 2019. "Improving non-hunters' attitudes toward hunting." *Human Dimensions of Wildlife* 24, no. 5: 480–487.

Boitani, Luigi, and Carlos Rondinini. 2010. "Mammals." In *Evolution Lost: Status and Trends of the World's Vertebrates*, edited by Jonathan E. M. Baillie, Janine Griffiths, Samuel T. Turvey, Jonathan Loh, and Ben Collen, 38–46. London: Zoological Society of London.

Bottrill, Madeleine C., Liana N. Joseph, Josie Carwardine, Michael Bode, Carly Cook, Edward T. Game, Hedley Grantham, et al. 2008. "Is conservation triage just smart decision making?" *Trends in Ecology & Evolution* 23, no. 12: 649–654.

Bourbeau-Lemieux, Aurélie, Marco Festa-Bianchet, Jean-Michel Gaillard, and Fanie Pelletier. 2011. "Predator-driven component Allee effects in a wild ungulate." *Ecology Letters* 14, no. 4: 358–363.

Bradshaw, Lauren, Robert H. Holsman, Jordan Petchenik, and Taylor Finger. 2019. "Meeting harvest expectations is key for duck hunter satisfaction." *Wildlife Society Bulletin* 43, no. 1: 102–111.

Branch, Trevor A., Aaron S. Lobo, and Steven W. Purcell. 2013. "Opportunistic exploitation: An overlooked pathway to extinction." *Trends in Ecology & Evolution* 28, no. 7: 409–413.

Brashares, Justin S., Jeffery R. Werner, and A. R. E. Sinclair. 2010. "Social 'meltdown' in the demise of an island endemic: Allee effects and the Vancouver Island marmot." *Journal of Animal Ecology* 79, no. 5: 965–973.

Braun, Martin. 1983. "Why the percentage of sharks caught in the Mediterranean Sea rose dramatically during World War I." In *Differential Equation Models*, edited by Martin Braun, Courtney S. Coleman, and Donald A. Drew, 221–228. New York: Springer.

Breisjøberget, Jo Inge, Morten Odden, Torstein Storaas, Erlend B. Nilsen, and Mikkel A. J. Kvasnes. 2018a. "Harvesting a red-listed species: Determinant factors for willow ptarmigan harvest rates, bag sizes, and hunting efforts in Norway." *European Journal of Wildlife Research* 64, no. 5: 1–10.

Breisjøberget, Jo Inge, Morten Odden, Per Wegge, Barbara Zimmermann, and Harry Andreassen. 2018b. "The alternative prey hypothesis revisited: Still valid for willow ptarmigan population dynamics." *PLoS ONE* 13, no. 6: e0197289.

Brodie, Jedediah, Heather Johnson, Michael Mitchell, Peter Zager, Kelly Proffitt, Mark Hebblewhite, Matthew Kauffman, et al. 2013. "Relative influence of human harvest, carnivores, and weather on adult female elk survival across western North America." *Journal of Applied Ecology* 50, no. 2: 295–305.

Brook, Barry W., Lochran W. Traill, and Corey J. A. Bradshaw. 2006. "Minimum viable population sizes and global extinction risk are unrelated." *Ecology Letters* 9, no. 4: 375–382.

Bruskotter, Jeremy T., John A. Vucetich, Sherry Enzler, Adrian Treves, and Michael P. Nelson. 2014. "Removing protections for wolves and the future of the US Endangered Species Act (1973)." *Conservation Letters* 7, no. 4: 401–407.

Bryant, Edwin H., Vickie L. Backus, Michael E. Clark, and David H. Reed. 1999. "Experimental tests of captive breeding for endangered species." *Conservation Biology* 13, no. 6: 1487–1496.

Burgman, Mark. 2004. "Expert frailties in conservation risk assessment and listing decisions." In *Threatened Species Legislation: Is It Just an Act?* edited by Pat Hutchings, Daniel Lunney and Chris Dickman, 20–29. Mosman, NSW, Australia: Royal Zoological Society of New South Wales.

Butler, Matthew J., Grant Harris, and Bradley N. Strobel. 2013. "Influence of whooping crane population dynamics on its recovery and management." *Biological Conservation* 162: 89–99.

Campbell-Staton, Shane C., Brian J. Arnold, Dominique Gonçalves, Petter Granli, Joyce Poole, Ryan A. Long, and Robert M. Pringle. 2021. "Ivory poaching and the rapid evolution of tusklessness in African elephants." *Science* 374, no. 6566: 483–487.

Carlson, Shelby C., John A. Vucetich, Lexi Galiardi, and Jeremy T. Bruskotter. 2023. "Attitudes toward cougar restoration in seven eastern states." https://doi.org/10.1101/2023.05.26.542322.

Carter, Sarah K., Glenn R. VanBlaricom, and Brian L. Allen. 2007. "Testing the generality of the trophic cascade paradigm for sea otters: A case study with kelp forests in northern Washington, USA." *Hydrobiologia* 579, no. 1: 233–249.

Catalano, Allison S., Joss Lyons-White, Morena M. Mills, and Andrew T. Knight. 2019. "Learning from published project failures in conservation." *Biological Conservation* 238: 108223.

Caughley, Graeme. 1994. "Directions in conservation biology." *Journal of Animal Ecology* 63, no. 2: 215–244.

Ceballos, Gerardo, and Paul R. Ehrlich. 2002. "Mammal population losses and the extinction crisis." *Science* 296, no. 5569: 904–907.

Ceballos, Gerardo, Paul R. Ehrlich, and Rodolfo Dirzo. 2017. "Biological annihilation via the ongoing sixth mass extinction signaled by vertebrate population losses and declines." *Proceedings of the National Academy of Sciences* 114, no. 30: E6089–E6096.

Chapron, Guillaume, Dale G. Miquelle, Amaury Lambert, John M. Goodrich, Stéphane Legendre, and Jean Clobert. 2008. "The impact on tigers of poaching versus prey depletion." *Journal of Applied Ecology* 45, no. 6: 1667–1674.

Charlesworth, Deborah, and John H. Willis. 2009. "The genetics of inbreeding depression." *Nature Reviews Genetics* 10, no. 11: 783–796.

Chaudhary, Vratika, and Madan K. Oli. 2020. "A critical appraisal of population viability analysis." *Conservation Biology* 34, no. 1: 26–40.

Clark, Colin W. 2010. *Mathematical Bioeconomics: The Mathematics of Conservation* (Vol. 91). Hoboken, NJ: John Wiley & Sons.

Clark, Francis, Barry W. Brook, Steven Delean, H. Reşit Akçakaya, and Corey J. A. Bradshaw. 2010. "The theta-logistic is unreliable for modelling most census data." *Methods in Ecology and Evolution* 1, no. 3: 253–262.

Clements, Gopalasamy Reuben, Corey J. A. Bradshaw, Barry W. Brook, and William F. Laurance. 2011. "The SAFE index: Using a threshold population target to measure relative species threat." *Frontiers in Ecology and the Environment* 9, no. 9: 521–525.

Clutton-Brock, Tim. 2007. "Sexual selection in males and females." *Science* 318, no. 5858: 1882–1885.

Coals, Peter, Dawn Burnham, Andrew Loveridge, David W. Macdonald, Michael 't Sas-Rolfes, Vivienne L. Williams, and John A. Vucetich. 2019. "The ethics of human–animal relationships and public discourse: A case study of lions bred for their bones." *Animals* 9, no. 2: 52.

Collen, Ben, Jonathan Loh, Sarah Whitmee, Louise McRae, Rajan Amin, and Jonathan E. M. Baillie. 2009. "Monitoring change in vertebrate abundance: The Living Planet Index." *Conservation Biology* 23, no. 2: 317–327.

Cooch, Evan G., Matthieu Guillemain, G. Scott Boomer, Jean-Dominique Lebreton, and James D. Nichols. 2014. "The effects of harvest on waterfowl populations." *Wildfowl* 4: 220–276.

Cooper, Andrew B., José C. Pinheiro, James W. Unsworth, and Ray Hilborn. 2002. "Predicting hunter success rates from elk and hunter abundance, season structure, and habitat." *Wildlife Society Bulletin* 30, no. 4: 1068–1077.

Courchamp, Franck, Rosie Woodroffe, and Gary Roemer. 2003. "Removing protected populations to save endangered species." *Science* 302, no. 5650: 1532–1532.

Cox, Michael, Gwen Arnold, and Sergio Villamayor Tomás. 2010. "A review of design principles for community-based natural resource management." *Ecology and Society* 15, no. 4: 38.

Crew, Bee. 2019. "The extraordinary survival story of the black robin." *Australian Geographic*, March 11. https://www.australiangeographic.com.au/blogs/creatura -blog/2019/03/the-extraordinary-survival-story-of-the-black-robin/.

Crnokrak, Peter, and Derek A. Roff. 1999. "Inbreeding depression in the wild." *Heredity* 83, no. 3: 260–270.

Darst, Catherine R., Philip J. Murphy, Nathan W. Strout, Steven P. Campbell, Kimberleigh J. Field, Linda Allison, and Roy C. Averill-Murray. 2013. "A strategy for prioritizing threats and recovery actions for at-risk species." *Environmental Management* 51, no. 3: 786–800.

Dawes, J. H. P., and M. O. Souza. 2013. "A derivation of Holling's type I, II and III functional responses in predator–prey systems." *Journal of Theoretical Biology* 327: 11–22.

Dean, Thomas A., James L. Bodkin, Stephen C. Jewett, Daniel H. Monson, and Dennis Jung. 2000. "Changes in sea urchins and kelp following a reduction in sea otter density as a result of the Exxon Valdez oil spill." *Marine Ecology Progress Series* 199: 281–291.

Devine, Jennifer A., Krista D. Baker, and Richard L. Haedrich. 2006. "Deep-sea fishes qualify as endangered." *Nature* 439, no. 7072: 29.

Dickson, Paul, and William M. Adams. 2009. "Science and uncertainty in South Africa's elephant culling debate." *Environment and Planning C: Government and Policy* 27, no. 1: 110–123.

Doak, Daniel F., Gina K. Boor, Victoria J. Bakker, William F. Morris, Allison Louthan, Scott A. Morrison, Amanda Stanley, et al. 2015. "Recommendations for improving recovery criteria under the US Endangered Species Act." *BioScience* 65, no. 2: 189–199.

Dobson, Andy, and Joyce Poole. 1998. "Conspecific aggregation and conservation biology." In *Behavioural Ecology and Conservation Biology*, edited by Tim Caro, 193–208. Oxford: Oxford University Press.

Dunk, Jeffrey R., Brian Woodbridge, Nathan Schumaker, Elizabeth M. Glenn, Brendan White, David W. LaPlante, Robert G. Anthony, et al. 2019. "Conservation planning for species recovery under the Endangered Species Act: A case study with the Northern Spotted Owl." *PLoS ONE* 14, no. 1: e0210643.

Durkin, Patrick. 2021. "Wisconsin's wolf overharvest was predictable, preventable." https://www.patrickdurkinoutdoors.com/post/wisconsin-s-wolf-overharvest -was-preventable.

Engen, Steinar, Russell Lande, and Bernt-Erik Sæther. 1997. "Harvesting strategies for fluctuating populations based on uncertain population estimates." *Journal of Theoretical Biology* 186, no. 2: 201–212.

Eriksen, Lasse F., Pål F. Moa, and Erlend B. Nilsen. 2018. "Quantifying risk of overharvest when implementation is uncertain." *Journal of Applied Ecology* 55, no. 2: 482–493.

Eriksson, Hampus, and Shelley Clarke. 2015. "Chinese market responses to overexploitation of sharks and sea cucumbers." *Biological Conservation* 184: 163–173.

Estes, James A., Charles H. Peterson, and Robert S. Steneck. 2010. "Some effects of apex predators in higher-latitude coastal oceans." In *Trophic Cascades Predators, Prey, and the Changing Dynamics of Nature*, edited by John Terborgh and James A. Estes, 37–53. Washington, DC: Island Press.

Fa, Julia E., Jesús Olivero, Raimundo Real, Miguel A. Farfán, Ana L. Márquez, J. Mario Vargas, Stefan Ziegler, et al. 2015. "Disentangling the relative effects of bushmeat availability on human nutrition in central Africa." *Scientific Reports* 5, no. 1: 1–8.

Fabiani, Anna, Filippo Galimberti, Simona Sanvito, and A. Rus Hoelzel. 2004. "Extreme polygyny among southern elephant seals on Sea Lion Island, Falkland Islands." *Behavioral Ecology* 15, no. 6: 961–969.

Fagan, William F., Eli Meir, John Prendergast, Ayoola Folarin, and Peter Karieva. 2001. "Characterizing population vulnerability for 758 species." *Ecology Letters* 4, no. 2: 132–138.

Fears, Daniel. 2016. "Red wolves will still be protected—but more by zoos than in the wild." *Washington Post*, September 12.

Ferreira, Sam M., Cathy Greaver, and Chenay Simms. 2017. "Elephant population growth in Kruger National Park, South Africa, under a landscape management approach." *Koedoe: African Protected Area Conservation and Science* 59, no. 1: 1–6.

Festa-Bianchet, Marco. 2003. "Exploitative wildlife management as a selective pressure for the life-history evolution of large mammals." In *Animal Behavior and Wildlife Conservation*, edited by Marco Festa-Bianchet and Marco Apollonio, 191–207. Washington, DC: Island Press.

Festa-Bianchet, Marco, and Atle Mysterud. 2018. "Hunting and evolution: Theory, evidence, and unknowns." *Journal of Mammalogy* 99, no. 6: 1281–1292.

Fieberg, John, and Stephen P. Ellner. 2001. "Stochastic matrix models for conservation and management: A comparative review of methods." *Ecology Letters* 4, no. 3: 244–266.

Finkelstein, Myra E., Daniel F. Doak, M. Nakagawa, P. R. Sievert, and J. Klavitter. 2010. "Assessment of demographic risk factors and management priorities: Impacts on juveniles substantially affect population viability of a long-lived seabird." *Animal Conservation* 13, no. 2: 148–156.

Food and Agriculture Organization of the United Nations (FAO). 2020. "Meat, total—Food available for consumption (kilograms per year per capita)," processed by Our World in Data. https://ourworldindata.org/grapher/daily-meat-consumption-per-person.

Food and Agriculture Organization of the United Nations (FAO). 2023. "FAOSTAT, Food balances (2010–)." https://www.fao.org/faostat/en/#data/FBS.

Forbes, Stephen H., and John T. Hogg. 1999. "Assessing population structure at high levels of differentiation: Microsatellite comparisons of bighorn sheep and large carnivores." *Animal Conservation* 2, no. 3: 223–233.

Fowler, Charles W. 1981. "Density dependence as related to life history strategy." *Ecology* 62, no. 3: 602–610.

Frankham, Richard. 1996. "Relationship of genetic variation to population size in wildlife." *Conservation Biology* 10, no. 6: 1500–1508.

Frankham, Richard. 2015. "Genetic rescue of small inbred populations: Meta-analysis reveals large and consistent benefits of gene flow." *Molecular Ecology* 24, no. 11: 2610–2618.

Frankham, Richard, Jonathan D. Ballou, and David A. Briscoe. 2002. *Introduction to Conservation Genetics.* Cambridge: Cambridge University Press.

Franklin, I. R. 1980. "Evolutionary change in small populations." In *Conservation Biology: An Evolutionary–Ecological Perspective,* edited by Michael E. Soulé and Bruce A. Wilcox, 135–150. Sunderland, MA: Sinauer.

Freddy, D. J. 1987. "The White River elk herd: A perspective 1960–85." *Wildlife Technical Publication* 37. Fort Collins: Colorado Division of Wildlife.

Fryxell, John M., Craig Packer, Kevin McCann, Erling J. Solberg, and Bernt-Erik Sæther. 2010. "Resource management cycles and the sustainability of harvested wildlife populations." *Science* 328, no. 5980: 903–906.

Fujiwara, Masami, and Hal Caswell. 2001. "Demography of the endangered North Atlantic right whale." *Nature* 414, no. 6863: 537–541.

Fuller Todd K., L. D. Mech, and J. F. Cochrane. 2003. "Wolf population dynamics." In *Wolves: Behavior, Ecology, and Conservation,* edited by L. D. Mech and Luigi Boitani, 161–91. Chicago: University of Chicago Press.

Gause, Georgii F. 1934. *The Struggle for Existence.* Baltimore, MD: Williams and Wilkins.

Gerber, Leah R. 2016. "Conservation triage or injurious neglect in endangered species recovery." *Proceedings of the National Academy of Sciences* 113, no. 13: 3563–3566.

Giles, Brian G., and C. Scott Findlay. 2004. "Effectiveness of a selective harvest system in regulating deer populations in Ontario." *Journal of Wildlife Management* 68, no. 2: 266–277.

Gilg, Olivier, Ilkka Hanski, and Benoît Sittler. 2003. "Cyclic dynamics in a simple vertebrate predator-prey community." *Science* 302, no. 5646: 866–868.

Gilg, Olivier, Benoît Sittler, Brigitte Sabard, Arnaud Hurstel, Raphaël Sané, Pierre Delattre, and Ilkka Hanski. 2006. "Functional and numerical responses of four lemming predators in high arctic Greenland." *Oikos* 113, no. 2: 193–216.

Ginzburg, Lev R. 1998. "Assuming reproduction to be a function of consumption raises doubts about some popular predator-prey models." *Journal of Animal Ecology* 67, no. 2: 325–327.

Golden, Christopher D., Lia C. H. Fernald, Justin S. Brashares, B. J. Rodolph Rasolofoniaina, and Claire Kremen. 2011. "Benefits of wildlife consumption to child nutrition in a biodiversity hotspot." *Proceedings of the National Academy of Sciences* 108, no. 49: 19653–19656.

Green, Samantha. 2018. "Why curiosity and wonder are critical for the next generation of scientists." The Wiley Network, January 17. https://www.wiley .com/en-us/network/publishing/societies/research-impact/why-curiosity-and -wonder-are-critical-for-the-next-generation-of-scientists.

Gross, Thilo, Lars Rudolf, Simon A. Levin, and Ulf Dieckmann. 2009. "Generalized models reveal stabilizing factors in food webs." *Science* 325, no. 5941: 747–750.

Grubbs, R. Dean, John K. Carlson, Jason G. Romine, Tobey H. Curtis, W. David McElroy, Camilla T. McCandless, Charles F. Cotton, et al. 2016. "Critical assessment and ramifications of a purported marine trophic cascade." *Scientific Reports* 6, no. 1: 1–12.

Gruen, Lori. 2021. "The moral status of animals." *The Stanford Encyclopedia of Philosophy* (summer 2021 edition), edited by Edward N. Zalta. https://plato .stanford.edu/archives/sum2021/entries/moral-animal/.

Haines, Aaron M., Michael E. Tewes, Linda L. Laack, William E. Grant, and John Young. 2005. "Evaluating recovery strategies for an ocelot (*Leopardus pardalis*) population in the United States." *Biological Conservation* 126, no. 4: 512–522.

Hall, Richard J., E. J. Milner-Gulland, and F. Courchamp. 2008. "Endangering the endangered: The effects of perceived rarity on species exploitation." *Conservation Letters* 1, no. 2: 75–81.

Hansson, Lennart. 1985. "Clethrionomys food: Generic, specific and regional characteristics." *Annales Zoologici Fennici* 22, no. 3: 315–318.

Harrington, Rhidian, Norman Owen-Smith, Petri C. Viljoen, Harry C. Biggs, Darryl R. Mason, and Paul Funston. 1999. "Establishing the causes of the roan antelope decline in the Kruger National Park, South Africa." *Biological Conservation* 90, no. 1: 69–78.

Harris, Michael. 2013. *Lament for an Ocean: The Collapse of the Atlantic Cod Fishery*. Toronto: McClelland & Stewart.

Hebblewhite, Mark, Marco Musiani, and L. Scott Mills. 2010. "Restoration of genetic connectivity among Northern Rockies wolf populations." *Molecular Ecology* 19, no. 20: 4383–4385.

Hedrick, Philip W., Marty Kardos, Rolf O. Peterson, and John A. Vucetich. 2017. "Genomic variation of inbreeding and ancestry in the remaining two Isle Royale wolves." *Journal of Heredity* 108, no. 2: 120–126.

Hedrick, Philip W., Rolf O. Peterson, Leah M. Vucetich, Jennifer R. Adams, and John A. Vucetich. 2014. "Genetic rescue in Isle Royale wolves: Genetic analysis and the collapse of the population." *Conservation Genetics* 15, no. 5: 1111–1121.

Hedrick, Philip W., Elaina M. Tuttle, and Rusty A. Gonser. 2018. "Negative-assortative mating in the white-throated sparrow." *Journal of Heredity* 109, no. 3: 223–231.

Heeren, Alexander, Gabriel Karns, Jeremy Bruskotter, Eric Toman, Robyn Wilson, and Harmony Szarek. 2017. "Expert judgment and uncertainty regarding the protection of imperiled species." *Conservation Biology* 31, no. 3: 657–665.

Heino, Mikko, Beatriz Díaz Pauli, and Ulf Dieckmann. 2015. "Fisheries-induced evolution." *Annual Review of Ecology, Evolution, and Systematics* 46: 461–480.

Heithaus, Michael R., Alejandro Frid, Aaron J. Wirsing, and Boris Worm. 2008. "Predicting ecological consequences of marine top predator declines." *Trends in Ecology & Evolution* 23, no. 4: 202–210.

Henden, J. A., R. A. Ims, N. G. Yoccoz, and S. T. Killengreen. 2011. "Declining willow ptarmigan populations: the role of habitat structure and community dynamics." *Basic and Applied Ecology* 12, no. 5: 413–422.

Hendry, James. 2021. "Elephant problem? No simple answers." *Africa Geographic Stories*, July 21. https://africageographic.com/stories/do-we-have-an-elephant-problem/.

Hoffmann, Michael, Jerrold L. Belant, Janice S. Chanson, Neil A. Cox, John Lamoreux, Ana S. L. Rodrigues, Jan Schipper, et al. 2011. "The changing fates of the world's mammals." *Philosophical Transactions of the Royal Society B: Biological Sciences* 366, no. 1578: 2598–2610.

Hoffmann, Michael, Craig Hilton-Taylor, Ariadne Angulo, Monika Böhm, Thomas M. Brooks, Stuart H. M. Butchart, Kent E. Carpenter, et al. 2010. "The impact of conservation on the status of the world's vertebrates." *Science* 330, no. 6010: 1503–1509.

Holling, Crawford S., and Gary K. Meffe. 1996. "Command and control and the pathology of natural resource management." *Conservation Biology* 10, no. 2: 328–337.

Holmes, Elizabeth E., John L. Sabo, Steven V. Viscido, and William F. Fagan. 2007. "A statistical approach to quasi-extinction forecasting." *Ecology Letters* 10, no. 12: 1182–1198.

Hörnfeldt, Birger. 1994. "Delayed density dependence as a determinant of vole cycles." *Ecology* 75, no. 3: 791–806.

Hoy, Sarah R, Kristin E. Brzeski, Leah M. Vucetich, Rolf O. Peterson, and John A. Vucetich. 2023. "The difficulty of detecting inbreeding depression and its effect on conservation decisions." *Journal of Heredity*, esad080. https://doi.org/10.1093/jhered/esad080.

Hoy, Sarah R., Daniel R. MacNulty, Matthew C. Metz, Douglas W. Smith, Daniel R. Stahler, Rolf O. Peterson, and John A. Vucetich. 2021. "Negative frequency-dependent prey selection by wolves and its implications on predator–prey dynamics." *Animal Behaviour* 179: 247–265.

Hoy, Sarah R., Daniel R. MacNulty, Douglas W. Smith, Daniel R. Stahler, Xavier Lambin, Rolf O. Peterson, Joel S. Ruprecht, and John A. Vucetich. 2020. "Fluctuations in age structure and their variable influence on population growth." *Functional Ecology* 34, no. 1: 203–216.

Hoy, Sarah R., Rolf O. Peterson, John A. Vucetich. 2022. *Ecological Studies of Wolves on Isle Royale Annual Report, 2021–2022*. Houghton: Michigan Technological University.

Hoy, Sarah R., Steve J. Petty, Alexandre Millon, D. Philip Whitfield, Michael Marquiss, D. I. K. Anderson, Martin Davison, and Xavier Lambin. 2017. "Density-dependent increase in superpredation linked to food limitation in a recovering population of northern goshawks, *Accipiter gentilis*." *Journal of Avian Biology* 48, no. 9: 1205–1215.

Hoy, Sarah R., Steve J. Petty, Alexandre Millon, D. Philip Whitfield, Michael Marquiss, Martin Davison, and Xavier Lambin. 2015. "Age and sex-selective predation moderate the overall impact of predators." *Journal of Animal Ecology* 84, no. 3: 692–701.

Huffaker, Carl Barton. 1958. "Experimental studies on predation: Dispersion factors and predator-prey oscillations." *Hilgardia* 27, no. 14: 343–383.

Hunter, Christine M., Hal Caswell, Michael C. Runge, Eric V. Regehr, Steve C. Amstrup, and Ian Stirling. 2010. "Climate change threatens polar bear populations: A stochastic demographic analysis." *Ecology* 91, no. 10: 2883–2897.

Jamieson, Ian G., and Fred W. Allendorf. 2012. "How does the 50/500 rule apply to MVPs?." *Trends in Ecology & Evolution* 27, no. 10: 578–584.

Jeschke, Jonathan M., Michael Kopp, and Ralph Tollrian. 2004. "Consumer-food systems: Why type I functional responses are exclusive to filter feeders." *Biological Reviews* 79, no. 2: 337–349.

Jiang, Yuexin, Daniel I. Bolnick, and Mark Kirkpatrick. 2013. "Assortative mating in animals." *American Naturalist* 181, no. 6: E125–E138.

Johnson, Warren E., David P. Onorato, Melody E. Roelke, E. Darrell Land, Mark Cunningham, Robert C. Belden, Roy McBride, et al. 2010. "Genetic restoration of the Florida panther." *Science* 329, no. 5999: 1641–1645.

Kahler, Jessica S., and Meredith L. Gore. 2012. "Beyond the cooking pot and pocket book: Factors influencing noncompliance with wildlife poaching rules." *International Journal of Comparative and Applied Criminal Justice* 36, no. 2: 103–120.

Kahneman, Daniel. 2011. *Thinking, Fast and Slow*. New York: Farrar, Straus, and Giroux.

Kareiva, Peter. 2001. "When one whale matters." *Nature* 414, no. 6863: 493–494.

Keller, Lukas F., and Amy B. Marr. 2006. *Conservation and Biology of Small Populations: The Song Sparrows of Mandarte Island*. Oxford: Oxford University Press.

Kendall, Bruce E., John Prendergast, and Ottar N. Bjørnstad. 1998. "The macroecology of population dynamics: Taxonomic and biogeographic patterns in population cycles." *Ecology Letters* 1, no. 3: 160–164.

Kenney, John, Fred W. Allendorf, Charles McDougal, and James L. D. Smith. 2014. "How much gene flow is needed to avoid inbreeding depression in wild tiger populations?" *Proceedings of the Royal Society B: Biological Sciences* 281, no. 1789: 20133337.

Klug, Fritz. 2014. "DNR calls first Michigan wolf hunt a 'success'; issue to continue to 2014 ballot." Mlive, January 1. https://www.mlive.com/news/2014/01/dnr _calls_first_michigan_wolf.html.

Knowlton, Frederick F., Eric M. Gese, and Michael M. Jaeger. 1999. "Coyote depredation control: An interface between biology and management." *Journal of Range Management* 52, no. 5: 398–412.

Kovach, Ryan P., Anthony J. Gharrett, and David A. Tallmon. 2012. "Genetic change for earlier migration timing in a pink salmon population." *Proceedings of the Royal Society B: Biological Sciences* 279, no. 1743: 3870–3878.

Krueger, Tobias, Trevor Page, Klaus Hubacek, Laurence Smith, and Kevin Hiscock. 2012. "The role of expert opinion in environmental modelling." *Environmental Modelling & Software* 36: 4–18.

Lacy, Robert C., Glen Alaks, and Allison Walsh. 1996. "Hierarchical analysis of inbreeding depression in *Peromyscus polionotus*." *Evolution* 50, no. 6: 2187–2200.

Lacy, Robert C., Philip S. Miller, and Kathy Traylor-Holzer. 2021. *Vortex 10 User's Manual, March 30, 2021 Update*. Apple Valley, MN: International Union for Conservation of Nature Species Survival Commission Conservation Planning Specialist Group and Chicago Zoological Society.

Lande, Russell. 1993. "Risks of population extinction from demographic and environmental stochasticity and random catastrophes." *American Naturalist* 142, no. 6: 911–927.

Lande, Russell. 1995. "Mutation and conservation." *Conservation Biology* 9, no. 4: 782–791.

Lande, Russell, Steinar Engen, and Bernt-Erik Sæther. 1995. "Optimal harvesting of fluctuating populations with a risk of extinction." *American Naturalist* 145, no. 5: 728–745.

Latta, Brian. 2001. "On California's Channel Islands, native predators became prey when feral pigs rearranged the food web." *Science Daily*, December 20. https://www.sciencedaily.com/releases/2001/12/011219062351.htm.

Latter, B. D., J. Cetal Mulley, D. Reid, and L. Pascoe. 1995. "Reduced genetic load revealed by slow inbreeding in *Drosophila melanogaster*." *Genetics* 139, no. 1: 287–297.

Lavers, Jennifer L., Chris Wilcox, and C. Josh Donlan. 2010. "Bird demographic responses to predator removal programs." *Biological Invasions* 12, no. 11: 3839–3859.

Lemos, Maria Carmen, and Arun Agrawal. 2006. "Environmental governance." *Annual Review of Environment and Resources* 31, no. 1: 297–325.

Lennox, Robert J., Austin J. Gallagher, Euan G. Ritchie, and Steven J. Cooke. 2018. "Evaluating the efficacy of predator removal in a conflict-prone world." *Biological Conservation* 224: 277–289.

Lewis, Sue, T. N. Sherratt, K. C. Hamer, and Sarah Wanless. 2001. "Evidence of intra-specific competition for food in a pelagic seabird." *Nature* 412, no. 6849: 816–819.

Lilley, Malin K., Kendal A. Smith, and Natalia Botero-Acosta. 2017. "Cetacean life history." In *Encyclopedia of Animal Cognition and Behavior*, edited by Jennifer Vonk and Todd Shackelford, 1–9. Springer. https://link.springer.com/referencework/10.1007/978-3-319-47829-6#bibliographic-information.

Lindenmayer, David B., Mark A. Burgman, H. R. Akçakaya, R. C. Lacy, and H. P. Possingham. 1995. "A review of the generic computer programs ALEX, RAMAS/space and VORTEX for modelling the viability of wildlife metapopulations." *Ecological Modelling* 82, no. 2: 161–174.

Lindsey, Peter A., Guillaume Chapron, Lisanne S. Petracca, Dawn Burnham, Matthew W. Hayward, Philipp Henschel, Amy E. Hinks, et al. 2017. "Relative efforts of countries to conserve world's megafauna." *Global Ecology and Conservation* 10: 243–252.

Lindström, Jan, Hanna Kokko, Esa Ranta, and Harto Lindén. 1999. "Density dependence and the response surface methodology." *Oikos* 85, no. 1: 40–52.

Lotka, Alfred James. 1925. *Elements of Physical Biology*. Baltimore, MD: Williams & Wilkins.

Luikart, Gordon, Nils Ryman, David A. Tallmon, Michael K. Schwartz, and Fred W. Allendorf. 2010. "Estimation of census and effective population sizes: The increasing usefulness of DNA-based approaches." *Conservation Genetics* 11, no. 2: 355–373.

Lynch, Michael, and Russell Lande. 1998. "The critical effective size for a genetically secure population." *Animal Conservation* 1, no. 1: 70–72.

Lyons, Jessica A., and Daniel J. D. Natusch. 2013. "Effects of consumer preferences for rarity on the harvest of wild populations within a species." *Ecological Economics* 93: 278–283.

Madsen, Thomas, Jon Loman, Lewis Anderberg, Håkan Anderberg, Arthur Georges, and Beata Ujvari. 2020. "Genetic rescue restores long-term viability of an isolated population of adders (*Vipera berus*)." *Current Biology* 30, no. 21: R1297–R1299.

Marshall, Kristin N., N. Thompson Hobbs, and David J. Cooper. 2013. "Stream hydrology limits recovery of riparian ecosystems after wolf reintroduction." *Proceedings of the Royal Society B: Biological Sciences* 280, no. 1756: 20122977.

Martin, Tara G., Mark A. Burgman, Fiona Fidler, Petra M. Kuhnert, Samantha Low-Choy, Marissa McBride, and Kerrie Mengersen. 2012. "Eliciting expert knowledge in conservation science." *Conservation Biology* 26, no. 1: 29–38.

Martin, Tara G., Laura Kehoe, Chrystal Mantyka-Pringle, Iadine Chades, Scott Wilson, Robin G. Bloom, Stephen K. Davis, Ryan Fisher, Jeff Keith, Katherine Mehl, and Beatriz Prieto Diaz. 2018. "Prioritizing recovery funding to maximize conservation of endangered species." *Conservation Letters* 11, no. 6: e12604.

Mason, Russ, Bill O'Neill, James Dexter, Gary Hagler, Ronald A. Olson, and William E. Moritz. 2013. Memorandum to the natural resources commission, May 13. Acc 16-022(8), Michigan Wolf Hunting Correspondence, Michigan Technological University Archives and Copper Country Historical Collections, Houghton.

Massaro, Melanie, Raazesh Sainudiin, Don Merton, James V. Briskie, Anthony M. Poole, and Marie L. Hale. 2013. "Human-assisted spread of a maladaptive behavior in a critically endangered bird." *PLoS ONE* 8, no. 12: e79066.

Maxwell, Sean L., Richard A. Fuller, Thomas M. Brooks, and James E. M. Watson. 2016. "Biodiversity: The ravages of guns, nets and bulldozers." *Nature* 536, no. 7615: 143–145.

McCann, Kevin. S. 2011. *Food Webs*. Princeton, NJ: Princeton University Press.

McCarthy, Michael A., Georgia E. Garrard, Alana L. Moore, Kirsten M. Parris, Tracey J. Regan, and Gerard E. Ryan. 2011. "The SAFE index should not be used for prioritization." *Frontiers in Ecology and the Environment* 9, no. 9: 486–487.

McCarthy, Michael A., David Keith, Justine Tietjen, Mark A. Burgman, Mark Maunder, Larry Master, Barry W. Brook, et al. 2004. "Comparing predictions of extinction risk using models and subjective judgement." *Acta Oecologica* 26, no. 2: 67–74.

McDonald-Madden, Eve, Peter W. J. Baxter, and Hugh P. Possingham. 2008. "Subpopulation triage: How to allocate conservation effort among populations." *Conservation Biology* 22, no. 3: 656–665.

McLellan, Bruce N., Robert Serrouya, Heiko U. Wittmer, and Stan Boutin. 2010. "Predator-mediated Allee effects in multi-prey systems." *Ecology* 91, no. 1: 286–292.

McManus, J. S., Amy J. Dickman, David Gaynor, B. H. Smuts, and David W. Macdonald. 2015. "Dead or alive? Comparing costs and benefits of lethal and non-lethal human–wildlife conflict mitigation on livestock farms." *Oryx* 49, no. 4: 687–695.

Michigan Department of Natural Resources. 2022. Michigan Wolf Management Plan, Updated 2015. MDNR, Wildlife Division Report No. 3604. https://www .michigan.gov/documents/dnr/wolf_management_plan_492568_7.pdf.

Miller, Sterling D., David K. Person, and R. Terry Bowyer. 2022. "Efficacy of killing large carnivores to enhance moose harvests: New insights from a long-term view." *Diversity* 14, no. 11: 939.

Mills, L. Scott, and Fred W. Allendorf. 1996. "The one-migrant-per-generation rule in conservation and management." *Conservation Biology* 10, no. 6: 1509–1518.

Milner, Jos M., Erlend B. Nilsen, and Harry P. Andreassen. 2007. "Demographic side effects of selective hunting in ungulates and carnivores." *Conservation Biology* 21, no. 1: 36–47.

Milner-Gulland, E. J., O. M. Bukreeva, T. Coulson, A. A. Lushchekina, M. V. Kholodova, A. B. Bekenov, and I. A. Grachev. 2003. "Reproductive collapse in saiga antelope harems." *Nature* 422, no. 6928: 135–135.

Milner-Gulland, E. J., K. Shea, H. Possingham, T. Coulson, and C. Wilcox. 2001. "Competing harvesting strategies in a simulated population under uncertainty." *Animal Conservation* 4, no. 2: 157–167.

Minnie, Liaan, Angela Gaylard, and Graham I. H. Kerley. 2016. "Compensatory life-history responses of a mesopredator may undermine carnivore management efforts." *Journal of Applied Ecology* 53, no. 2: 379–387.

Miquelle, Dale G., Evgeny N. Smirnov, Olga Yu Zaumyslova, Svetlana V. Soutyrina, and Douglas H. Johnson. 2015. "Population dynamics of Amur tigers (*Panthera tigris altaica*) in Sikhote-Alin Biosphere Zapovednik: 1966–2012." *Integrative Zoology* 10, no. 4: 315–328.

Montgomery, Robert A. 2020. "Poaching is not one big thing." *Trends in Ecology & Evolution* 35, no. 6: 472–475.

Moss, Cynthia J. 2001. "The demography of an African elephant (*Loxodonta africana*) population in Amboseli, Kenya." *Journal of Zoology* 255, no. 2: 145–156.

Mudumba, Tutilo, Remington J. Moll, Sophia Jingo, Shawn Riley, David W. Macdonald, Christos Astaras, and Robert A. Montgomery. 2022. "Influence of social status and industrial development on poaching acceptability." *Global Ecology and Conservation* 38: e02271.

Muth, Robert M., and John F. Bowe Jr. 1998. "Illegal harvest of renewable natural resources in North America: Toward a typology of the motivations for poaching." *Society & Natural Resources* 11, no. 1: 9–24.

Nattrass, Stuart, Darren P. Croft, Samuel Ellis, Michael A. Cant, Michael N. Weiss, Brianna M. Wright, Eva Stredulinsky, Thomas Doniol-Valcroze, John K. Ford, Kenneth C. Balcomb, and Daniel W. Franks. 2019. "Postreproductive killer whale grandmothers improve the survival of their grandoffspring." *Proceedings of the National Academy of Sciences* 116, no. 52: 26669–26673.

Nunney, Leonard. 1996. "The influence of variation in female fecundity on effective population size." *Biological Journal of the Linnean Society* 59, no. 4: 411–425.

Odum, Eugene Pleasants, and Gary W. Barrett. 1971. *Fundamentals of Ecology*, vol. 3. Philadelphia: Saunders.

O'Grady, Julian J., Barry W. Brook, David H. Reed, Jonathan D. Ballou, David W. Tonkyn, and Richard Frankham. 2006. "Realistic levels of inbreeding depression strongly affect extinction risk in wild populations." *Biological Conservation* 133, no. 1: 42–51.

Owen-Smith, N., G. I. H. Kerley, B. Page, R. Slotow, and R. J. van Aarde. 2006. "A scientific perspective on the management of elephants in the Kruger National Park and elsewhere: Elephant conservation." *South African Journal of Science* 102, no. 9: 389–394.

Packer, Craig, Robert D. Holt, Peter J. Hudson, Kevin D. Lafferty, and Andrew P. Dobson. 2003. "Keeping the herds healthy and alert: Implications of predator control for infectious disease." *Ecology Letters* 6, no. 9: 797–802.

Packer, Craig, Marc Tatar, and Anthony Collins. 1998. "Reproductive cessation in female mammals." *Nature* 392, no. 6678: 807–811.

Palmer, William E., Shane D. Wellendorf, James R. Gillis, and Peter T. Bromley. 2005. "Effect of field borders and nest-predator reduction on abundance of northern bobwhites." *Wildlife Society Bulletin* 33, no. 4: 1398–1405.

Parkes, John P., David S. L. Ramsey, Norman Macdonald, Kelvin Walker, Sean McKnight, Brian S. Cohen, and Scott A. Morrison. 2010. "Rapid eradication of feral pigs (*Sus scrofa*) from Santa Cruz Island, California." *Biological Conservation* 143, no. 3: 634–641.

Pelletier, Fanie, John T. Hogg, and Marco Festa-Bianchet. 2006. "Male mating effort in a polygynous ungulate." *Behavioral Ecology and Sociobiology* 60, no. 5: 645–654.

Peterson, Rolf O., John A. Vucetich, Joseph M. Bump, and Douglas W. Smith. 2014. "Trophic cascades in a multicausal world: Isle Royale and Yellowstone." *Annual Review of Ecology, Evolution, and Systematics* 45: 325–345.

Pimentel, David. 1963. "Introducing parasites and predators to control native pests." *Canadian Entomologist* 95, no. 8: 785–792.

Pimm, Stuart L., Clinton N. Jenkins, Robin Abell, Thomas M. Brooks, John L. Gittleman, Lucas N. Joppa, Peter H. Raven, Callum M. Roberts, and Joseph O. Sexton. 2014. "The biodiversity of species and their rates of extinction, distribution, and protection." *Science* 344, no. 6187: 1246752.

Podt, Annemieke. 2017. "Orca Granny: Was she really 105?" *Orcazine*, January. http://orcazine.com/granny-j2/.

Pope, Karen L., Justin M. Garwood, Hartwell H. Welsh Jr, and Sharon P. Lawler. 2008. "Evidence of indirect impacts of introduced trout on native amphibians via facilitation of a shared predator." *Biological Conservation* 141, no. 5: 1321–1331.

Pöysä, Hannu, Lisa Dessborn, Johan Elmberg, Gunnar Gunnarsson, Petri Nummi, Kjell Sjöberg, Sari Suhonen, and Pär Söderquist. 2013. "Harvest mortality in North American mallards: A reply to Sedinger and Herzog." *Journal of Wildlife Management* 77, no. 4: 653–654.

Pöysä, H., J. Elmberg, G. Gunnarsson, P. Nummi, and K. Sjöberg. 2004. "Ecological basis of sustainable harvesting: Is the prevailing paradigm of compensatory mortality still valid?" *Oikos* 104, no. 3: 612–615.

Räikkönen, Jannikke, John A. Vucetich, Rolf O. Peterson, and Michael P. Nelson. 2009. "Congenital bone deformities and the inbred wolves (*Canis lupus*) of Isle Royale." *Biological Conservation* 142, no. 5: 1025–1031.

Ralls, Katherine, and Jonathan Ballou. 1983. "Extinction: Lessons from zoos." In *Genetics and Conservation: A Reference for Managing Wild Animal and Plant Populations*, edited by Christine M. Schonewald-Cox, Steven M. Chambers, Bruce MacBryde, and Lawrence Thomas, 164–183. Menlo Park, CA: Benjamin Cummings.

Ralls, Katherine, Jonathan D. Ballou, and Alan Templeton. 1988. "Estimates of lethal equivalents and the cost of inbreeding in mammals." *Conservation Biology* 2, no. 2: 185–193.

Réale, Denis, Andrew G. McAdam, Stan Boutin, and Dominique Berteaux. 2003. "Genetic and plastic responses of a northern mammal to climate change." *Proceedings of the Royal Society of London. Series B: Biological Sciences* 270, no. 1515: 591–596.

Reed, David H., and Edwin H. Bryant. 2000. "Experimental tests of minimum viable population size." *Animal Conservation* 3, no. 1: 7–14.

Reed, David H., Julian J. O'Grady, Jonathan D. Ballou, and Richard Frankham. 2003a. "The frequency and severity of catastrophic die-offs in vertebrates." *Animal Conservation* 6, no. 2: 109–114.

Reed, David H., Julian J. O'Grady, Barry W. Brook, Jonathan D. Ballou, and Richard Frankham. 2003b. "Estimates of minimum viable population sizes for vertebrates and factors influencing those estimates." *Biological Conservation* 113, no. 1: 23–34.

Reid, Jane M., Peter Arcese, Alice L. E. V. Cassidy, Amy B. Marr, James N. M. Smith, and Lukas F. Keller. 2005. "Hamilton and Zuk meet heterozygosity? Song repertoire size indicates inbreeding and immunity in song sparrows (*Melospiza melodia*)." *Proceedings of the Royal Society B: Biological Sciences* 272, no. 1562: 481–487.

Ripa, Jörgen, and Per Lundberg. 1996. "Noise colour and the risk of population extinctions." *Proceedings of the Royal Society of London. Series B: Biological Sciences* 263, no. 1377: 1751–1753.

Ripple, William J., James A. Estes, Robert L. Beschta, Christopher C. Wilmers, Euan G. Ritchie, Mark Hebblewhite, Joel Berger, et al. 2014. "Status and ecological effects of the world's largest carnivores." *Science* 343, no. 6167: 1241484.

Roblin, Sébastien. 2021. "The Air Force admits the F-35 fighter jet costs too much." NBC News, March 7. https://www.nbcnews.com/think/opinion/air-force-admits-f-35-fighter-jet-costs-too-much-ncna1259781.

Rodenhouse, Nicholas L., T. Scott Sillett, Patrick J. Doran, and Richard T. Holmes. 2003. "Multiple density-dependence mechanisms regulate a migratory bird population during the breeding season." *Proceedings of the Royal Society of London. Series B: Biological Sciences* 270, no. 1529: 2105–2110.

Rodrigues, Ana S. L., Thomas M. Brooks, Stuart H. M. Butchart, Janice Chanson, Neil Cox, Michael Hoffmann, and Simon N. Stuart. 2014. "Spatially explicit trends in the global conservation status of vertebrates." *PLoS ONE* 9, no. 11: e113934.

Roemer, Gary W., C. Josh Donlan, and Franck Courchamp. 2002. "Golden eagles, feral pigs, and insular carnivores: How exotic species turn native predators into prey." *Proceedings of the National Academy of Sciences* 99, no. 2: 791–796.

Rowlatt, Justin. 2022. "Scientists design contraceptives to limit grey squirrels." BBC News, July 11. https://www.bbc.com/news/science-environment-62096272.

Royama, Tomo. 2012. *Analytical Population Dynamics*. New York: Springer.

Saccheri, Ilik, Mikko Kuussaari, Maaria Kankare, Pia Vikman, Wilhelm Fortelius, and Ilkka Hanski. 1998. "Inbreeding and extinction in a butterfly metapopulation." *Nature* 392, no. 6675: 491–494.

Sæther, Bernt-Erik, and Øyvind Bakke. 2000. "Avian life history variation and contribution of demographic traits to the population growth rate." *Ecology* 81, no. 3: 642–653.

Sæther, Bernt-Erik, Steinar Engen, Flurin Filli, Ronny Aanes, Wolfgang Schröder, and Reidar Andersen. 2002. "Stochastic population dynamics of an introduced Swiss population of the ibex." *Ecology* 83, no. 12: 3457–3465.

Salo, Pälvi, Peter B. Banks, Chris R. Dickman, and Erkki Korpimäki. 2010. "Predator manipulation experiments: Impacts on populations of terrestrial vertebrate prey." *Ecological Monographs* 80, no. 4: 531–546.

Sandercock, Brett K., Erlend B. Nilsen, Henrik Brøseth, and Hans C. Pedersen. 2011. "Is hunting mortality additive or compensatory to natural mortality? Effects of experimental harvest on the survival and cause-specific mortality of willow ptarmigan." *Journal of Animal Ecology* 80, no. 1: 244–258.

Schmitz, Oswald J. 2004. "Perturbation and abrupt shift in trophic control of biodiversity and productivity." *Ecology Letters* 7, no. 5: 403–409.

Schmitz, Oswald J., Peter A. Hambäck, and Andrew P. Beckerman. 2000. "Trophic cascades in terrestrial systems: A review of the effects of carnivore removals on plants." *American Naturalist* 155, no. 2: 141–153.

Schneider, Richard R., Grant Hauer, and Stan Boutin. 2010. "Triage for conserving populations of threatened species: The case of woodland caribou in Alberta." *Biological Conservation* 143, no. 7: 1603–1611.

Schumaker, Nathan H., and Allen Brookes. 2018. "HexSim: A modeling environment for ecology and conservation." *Landscape Ecology* 33, no. 2: 197–211.

Sergio, Fabrizio, Luigi Marchesi, and Paolo Pedrini. 2003. "Spatial refugia and the coexistence of a diurnal raptor with its intraguild owl predator." *Journal of Animal Ecology* 72, no. 2: 232–245.

Sibly, Richard M., Daniel Barker, Jim Hone, and Mark Pagel. 2007. "On the stability of populations of mammals, birds, fish and insects." *Ecology Letters* 10, no. 10: 970–976.

Sillett, T. Scott, Nicholas L. Rodenhouse, and Richard T. Holmes. 2004. "Experimentally reducing neighbor density affects reproduction and behavior of a migratory songbird." *Ecology* 85, no. 9: 2467–2477.

Sinclair, Anthony R. E., Simon Mduma, and Justin S. Brashares. 2003. "Patterns of predation in a diverse predator–prey system." *Nature* 425, no. 6955: 288–290.

Shaffer, Mark L. 1981. "Minimum population sizes for species conservation." *BioScience* 31, no. 2: 131–134.

Skalski, Garrick T., and James F. Gilliam. 2001. "Functional responses with predator interference: Viable alternatives to the Holling type II model." *Ecology* 82, no. 11: 3083–3092.

Sneath, Sara. 2018. "14 endangered whooping cranes may be moved from Florida to Louisiana." Nola, March 15. https://www.nola.com/news/environment/article_c0e5944d-f951-5598-ac2b-23d0515fbc2f.html.

Solberg, Erling J., Anne Loison, Tor H. Ringsby, Bernt-Erik Sæther, and Morten Heim. 2002. "Biased adult sex ratio can affect fecundity in primiparous moose *Alces alces*." *Wildlife Biology* 8, no. 2: 117–128.

Soulé, Michael. E. 1976. "Allozyme variation, its determinants in space and time." In *Molecular Evolution*, edited by F. J. Ayala, 60–77 Sunderland, MA: Sinauer.

Staples, David F., Mark L. Taper, and Brian Dennis. 2004. "Estimating population trend and process variation for PVA in the presence of sampling error." *Ecology* 85, no. 4: 923–929.

Stone, Suzanne A., Stewart W. Breck, Jesse Timberlake, Peter M. Haswell, Fernando Najera, Brian S. Bean, and Daniel J. Thornhill. 2017. "Adaptive use of nonlethal strategies for minimizing wolf–sheep conflict in Idaho." *Journal of Mammalogy* 98, no. 1: 33–44.

Sustainable Fisheries. 2022. "What does the world eat?" https://sustainablefisheries-uw.org/seafood-101/what-does-the-world-eat/.

Sutherland, William J. 2001. "Sustainable exploitation: A review of principles and methods." *Wildlife Biology* 7, no. 3: 131–140.

Swain, Douglas P., and Hugues P. Benoît. 2015. "Extreme increases in natural mortality prevent recovery of collapsed fish populations in a Northwest Atlantic ecosystem." *Marine Ecology Progress Series* 519: 165–182.

Taleb, Nassim Nicholas. 2007. *The Black Swan: The Impact of the Highly Improbable*. New York: Random House.

Thompson, Grant G. 1991. "Determining minimum viable populations under the Endangered Species Act." *NOAA Technical memoir NMFS-F/NWC-198*, Washington, DC: US Department of Commerce.

Trask, Amanda E., Eric M. Bignal, Davy I. McCracken, Pat Monaghan, Stuart B. Piertney, and Jane M. Reid. 2016. "Evidence of the phenotypic expression of a lethal recessive allele under inbreeding in a wild population of conservation concern." *Journal of Animal Ecology* 85, no. 4: 879–891.

Turchin, Peter, and Ilkka Hanski. 1997. "An empirically based model for latitudinal gradient in vole population dynamics." *American Naturalist* 149, no. 5: 842–874.

Turkalo, Andrea K., Peter H. Wrege, and George Wittemyer. 2017. "Slow intrinsic growth rate in forest elephants indicates recovery from poaching will require decades." *Journal of Applied Ecology* 54, no. 1: 153–159.

Twining, Joshua P., W. Ian Montgomery, and David G. Tosh. 2021. "Declining invasive grey squirrel populations may persist in refugia as native predator recovery reverses squirrel species replacement." *Journal of Applied Ecology* 58, no. 2: 248–260.

US Fish and Wildlife Service. 2002. "Steller's eider recovery plan." Fairbanks, AL.

US Fish and Wildlife Service. 2009. "Revised recovery plan for the 'Alala (*Corvus hawaiiensis*)." Portland, OR.

US Fish and Wildlife Service. 2016. "Memorandum to the regional director, southeast region. RE: Recommended decisions in response to Red Wolf Recovery Program evaluation." https://www.fws.gov/redwolf/docs/recommended-decisions-in-response-to-red-wolf-recovery-program-evaluation.pdf.

US Fish and Wildlife Service. 2018. "Whooping crane survey results: Winter 2017–2018." https://www.fws.gov/uploadedFiles/WHCR%20Update%20Winter%202017-2018(1).pdf.

Utida, Syunro. 1957. "Cyclic fluctuation of population density intrinsic to the host–parasite system." *Ecology* 38, no. 3: 442–449.

van Aarde, Rudi, Ian Whyte, and Stuart Pimm. 1999. "Culling and the dynamics of the Kruger National Park African elephant population." *Animal Conservation* 2, no. 4: 287–294.

van Leeuwen, A., A. M. De Roos, and Lennart Persson. 2008. "How cod shapes its world." *Journal of Sea Research* 60, no. 1–2: 89–104.

Vila, Carles, Anna–Karin Sundqvist, Øystein Flagstad, Jennifer Seddon, Susanne Björnerfeldt, Ilpo Kojola, Adriano Casulli, Håkan Sand, Petter Wabakken, and Hans Ellegren. 2003. "Rescue of a severely bottlenecked wolf (*Canis lupus*) population by a single immigrant." *Proceedings of the Royal Society of London. Series B: Biological Sciences* 270, no. 1510: 91–97.

Volterra, Vito. 1926. "Fluctuations in the abundance of a species considered mathematically." *Nature* 118, no. 2972: 558–560.

Vucetich, John A. 2021. *Restoring the Balance: What Wolves Tell Us about Our Relationship with Nature.* Baltimore, MD: Johns Hopkins University Press.

Vucetich, John A., Jeremy T. Bruskotter, Nicholas Arrivo, and Mike Phillips. 2024. "A proposed policy for interpreting 'significant portion of its range' for the U.S. Endangered Species Act, 1973." *Georgetown Environmental Law Review*, in press.

Vucetich, John A., Jeremy T. Bruskotter, Michael Paul Nelson, Rolf O. Peterson, and Joseph K. Bump. 2017. "Evaluating the principles of wildlife conservation: A case study of wolf (*Canis lupus*) hunting in Michigan, United States." *Journal of Mammalogy* 98, no. 1: 53–64.

Vucetich, John A., Dawn Burnham, Paul J. Johnson, Andrew J. Loveridge, Michael Paul Nelson, Jeremy T. Bruskotter, and David W. Macdonald. 2019. "The value of argument analysis for understanding ethical considerations pertaining to trophy hunting and lion conservation." *Biological Conservation* 235: 260–272.

Vucetich, John A., Richard Damania, Sam A. Cushman, Ewan A. Macdonald, Dawn Burnham, Thomas Offer-Westort, Jeremy T. Bruskotter, Adam Feltz, Lily Van Eeden, and David W. Macdonald. 2021. "A minimally nonanthropocentric economics: What is it, is it necessary, and can it avert the biodiversity crisis?" *BioScience* 71, no. 8: 861–873.

Vucetich, John A., Mark Hebblewhite, Douglas W. Smith, and Rolf O. Peterson. 2011. "Predicting prey population dynamics from kill rate, predation rate and predator–prey ratios in three wolf-ungulate systems." *Journal of Animal Ecology* 80, no. 6: 1236–1245.

Vucetich, John. A., and Michael Paul Nelson. 2012. *A Handbook of Conservation and Sustainability Ethics.* Conservation Ethics Group, Occasional Paper Series 1. Self-pub., CreateSpace.

Vucetich, John A., and Michael Paul Nelson. 2018. "Acceptable risk of extinction in the context of endangered species policy." In *Philosophy and Public Policy*, edited by Andrew. I. Cohen, 81–104. New York: Rowman & Littlefield.

Vucetich, John A., and Rolf O. Peterson. 2004. "The influence of prey consumption and demographic stochasticity on population growth rate of Isle Royale wolves *Canis lupus.*" *Oikos* 107, no. 2: 309–320.

Vucetich, John A., Rolf O. Peterson, and Carrie L. Schaefer. 2002. "The effect of prey and predator densities on wolf predation." *Ecology* 83, no. 11: 3003–3013.

Vucetich, John A., Rolf O. Peterson, and Thomas A. Waite. 2004. "Raven scavenging favours group foraging in wolves." *Animal Behaviour* 67, no. 6: 1117–1126.

Vucetich, John A., Douglas W. Smith, and Daniel R. Stahler. 2005 "Influence of harvest, climate and wolf predation on Yellowstone elk, 1961-2004." *Oikos* 111, no. 2: 259–270.

Vucetich, John A., and Thomas A. Waite. 1998. "On the interpretation and application of mean times to extinction." *Biodiversity & Conservation* 7, no. 12: 1539–1547.

Vucetich, John A., and Thomas A. Waite. 1999. "Erosion of heterozygosity in fluctuating populations." *Conservation Biology* 13, no. 4: 860–868.

Vucetich, John A., Leah M. Vucetich, and Rolf O. Peterson. 2012. "The causes and consequences of partial prey consumption by wolves preying on moose." *Behavioral Ecology and Sociobiology* 66, no. 2: 295–303.

Wade, Paul R., Randall R. Reeves, and Sarah L. Mesnick. 2012. "Social and behavioural factors in cetacean responses to overexploitation: Are odontocetes less 'resilient' than mysticetes?" *Journal of Marine Biology* 2012, 567276.

Walters, Carl, and James F. Kitchell. 2001. "Cultivation/depensation effects on juvenile survival and recruitment: Implications for the theory of fishing." *Canadian Journal of Fisheries and Aquatic Sciences* 58, no. 1: 39–50.

Wcisel, Michelle, M. Justin O'Riain, Alta de Vos, and Wilfred Chivell. 2015. "The role of refugia in reducing predation risk for Cape fur seals by white sharks." *Behavioral Ecology and Sociobiology* 69, no. 1: 127–138.

Westemeier, Ronald L., Jeffrey D. Brawn, Scott A. Simpson, Terry L. Esker, Roger W. Jansen, Jeffery W. Walk, Eric L. Kershner, Juan L. Bouzat, and Ken N. Paige. 1998. "Tracking the long-term decline and recovery of an isolated population." *Science* 282, no. 5394: 1695–1698.

Whiteley, Andrew R., Sarah W. Fitzpatrick, W. Chris Funk, and David A. Tallmon. 2015. "Genetic rescue to the rescue." *Trends in Ecology & Evolution* 30, no. 1: 42–49.

Whyte, Ian J. 2004. "Ecological basis of the new elephant management policy for Kruger National Park and expected outcomes." *Pachyderm* 36: 99–108.

Whyte, Ian J. 2007. "Census results for elephant and buffalo in the Kruger National Park in 2006 and 2007." Internal Scientific Report No. 06/07. South African National Parks.

Whyte, Ian J., H. C. Biggs, A. Gaylard, and L. E. O. Braack. 1999. "A new policy for the management of the Kruger National Park's elephant population." *Koedoe* 42, no. 1: 111–132.

Whyte, Ian, Rudi van Aarde, and Stuart L. Pimm. 1998. "Managing the elephants of Kruger national park." *Animal Conservation* 1, no. 2: 77–83.

Whyte, Ian J., Rudi J. van Aarde, and Stuart L. Pimm. 2003. "Kruger's elephant population: Its size and consequences for ecosystem heterogeneity." In *The Kruger Experience: Ecology and Manangement of Savanna Heterogniety*, edited by Johan T. du Toit, Kevin H. Rogers, and Harry C. Biggs, 332–348. Washington, DC: Island Press.

Williams, Vivienne L. 2015. "Tiger-bone trade could threaten lions." *Nature* 523, no. 7560: 290–290.

Wilmers, Christopher C., Eric Post, and Alan Hastings. 2007. "A perfect storm: The combined effects on population fluctuations of autocorrelated environmental noise, age structure, and density dependence." *American Naturalist* 169, no. 5: 673–683.

Wittemyer, George, David Daballen, and Iain Douglas-Hamilton. 2013. "Comparative demography of an at-risk African elephant population." *PLoS ONE* 8, no. 1: e53726.

World Wide Fund for Nature (WWF). 2022. *Living Planet Report 2022—Building a Nature-Positive Society*, edited by R. E. A. Almond, M. Grooten, D. Juffe Bignoli, and T. Petersen. Gland, Switzerland.

Wright, Sewall. 1931. "Evolution in Mendelian populations." *Genetics* 16, no. 2: 97–159.

Yiming, L. I., and David S. Wilcove. 2005. "Threats to vertebrate species in China and the United States." *BioScience* 55, no. 2: 147–153.

Yu, Jinping, and F. Stephen Dobson. 2000. "Seven forms of rarity in mammals." *Journal of Biogeography* 27, no. 1: 131–139.

Zanette, Liana Y., and Michael Clinchy. 2020. "Ecology and neurobiology of fear in free-living wildlife." *Annual Review of Ecology, Evolution, and Systematics* 51, no. 1: 297–318.

INDEX

Page numbers followed by *f* and *t* indicate figures and tables.

Arabian oryx, inbreeding depression in, 150

arctic fox, predation by, 233–34

argument analysis: building, 51–54; building of arguments in, 51–54; confirmation bias as obstacle to, 53–54; critical thinking and, 49; definition of, 46, 47–49; evaluating premises of, 54–56; evaluation of premises in, 54–56; examples of, 46–49; goal of, 63–64; logical fallacies in, 47–49, 250n1; misperceptions about, 49–50; premises in, 47–49. *See also* culling of elephants, Kruger National Park

Asia, bushmeat consumption in, 169

assortative mating, 134

attack rate, 210, 214, 216

baboon(s), age-specific vital rates for, 68f

balancing selection, 132–33

bald eagle, 256n16

baleen whale, 197–98

Ballou, Jonathon, 146–47

bank vole: gray-sided vole compared to, 41; nonlinear density dependence, 33–34, 34f

bass, resiliency to overexploitation, 198–99

beach mouse, inbreeding depression in, 151, 152f

bean beetle, adzuki, predation and, 218f, 219, 220

bear, grizzly: impact on elk population, 239; predator control of, 244–45; trophy hunting of, 204

bear, polar, structured dynamics of, 91–93, 92f, 95

Beaufort Sea, polar bear population in, 91–93, 92f

beetle, adzuki bean (*Callosobruchus chinensis*), predation on, 218f, 219, 220

beluga(s), resiliency to overexploitation, 197–98

bias, confirmation, 53–54

bighorn sheep: density-dependent dynamics of, 39–40, 40f; exploitation of, 195; genetic rescue of, 165

biodiversity crisis: components of, 1–2; conservation triage and, 112–21; extinction rate and, 122–24; hope and fatalism and, 126–28, 253n24; impact of elephant culling on, 50–51, 57; seriousness of, 94–95. *See also* extinction risk

bio-economics, exploited populations and, 186–87

birds: black grouse, 44; black robin, 137–38; black-throated blue warbler, 27–33, 27f; burrowing owl, 116; condor, 137; glaucous-winged gull, 237; greater sage grouse, 116; great reed warbler, 150; great tit, 82–86, 82f, 85f, 86f; Hawaiian crow, 95; Laysan albatross, 86–88, 88f; long-tailed skua, 233–34; moorhen, 157–58; northern fulmar, 82–86, 82f, 83t, 86f; northern gannet, 25–26, 26f; northern spotted owl, 109, 110–11; personhood of, 249n1; pine marten, 220; population viability analysis (PVA) of, 103f; prairie chicken, 165, 165f; range contraction of, 1–2; red-billed chough, 135f, 136–38; red-cockaded woodpecker, 110, 150; snail kite, 34f; snowy owl, 233–34; song sparrow, 148–51, 149f; tawny owl, 220, 225–26; white-throated sparrow, 134; whooping crane, 17, 18f, 19t, 21–23, 23f, 45, 110; willow ptarmigan (*see* ptarmigan, willow, exploitation of)

bison, European, 253n4

black bear, predator control of, 245

black jackal, compensatory mortality in, 195

black robin, 137–38

black swan events, 246, 257n32

black-throated blue warbler, density-independent population growth, 27–33, 27f

Blake, William, 206
blue whale, 197–98
Botswana, ban on elephant hunting in, 13f
bottom-up predation processes, 228, 242–43, 242f
budmoth, larch, 44
burden of proof, 61
burrowing owl, 116
bushmeat, 169, 204
butterfly, Glanville fritillary, inbreeding depression in, 154–55, 155f

Callosobruchus chinensis. See beetle, adzuki bean (*Callosobruchus chinensis*)
Canada: lynx and snowshoe hares of, 220–21, 221f; Milk River Watershed, 116–17; red squirrels of Kluane National Park, 133; song sparrows of Mandarte Island, 148–51, 149f; whooping cranes in, 17; woodland caribou in, 115–16
caribou: apparent competition and, 231–32; conservation triage of, 115–16; selective exploitation of, 193. *See also* ungulate populations, exploitation of
cascade frog, apparent competition and, 232
cell-mediated immune response, in Mandarte Island song sparrows, 148–49
Channel Islands, polyphagous predation in, 230–31, 256n16
Chatham Islands, black robins in, 137–38
Chaudhary, Vratika, 106
chicken, prairie, genetic rescue of, 165, 165f
China: overexploitation in, 169–70; saiga antelope horn trade in, 192
Chippewa Harbor Pack wolves, Isle Royale National Park, 3–4
chondrodystrophy, 137
chough, red-billed, lethal alleles in, 135f, 136–38

ciliated protozoans, predation and, 215–17, 218f, 220, 255n5
climate change: ethical arguments about, 47; hope and fatalism in, 128; larch budmoth population cycles and, 44; pink salmon migration and, 132; rodent cycles and, 172
Coals, Peter, 204
cod, overexploitation of, 198, 200
collared lemming, impact of predation on, 233–34
collective action, 187
column vectors, structured populations, 75–78
communities, definition of, 5
compensatory mortality, 194–95, 194f
competition: apparent, 231–32, 253n15; contest, 256n14; exploitive, 252n13; interference, 252n13; scramble, 256n14
condor, dwarfism in, 137
confirmation bias, 53–54
consequentialism, 128, 250n4
conservation decision-making: adaptive management, 62–63, 62f; ethical inquiry and, 62–63
conservation ethics. *See* ethics
conservation genetics, 132. *See also* genetic diversity
conservation triage: algorithms for, 117–18; concerns about, 117–18; controversial nature of, 121–26; defined, 112–13; faux triage, 119; of grassland ecosystem, 116–17; of kit foxes, 113–14; miscommunication about, 120; scale or scope of, 118; *Sophie's Choice* analogy for, 120; strategic weaknesses of, 118–19; of Sumatran tiger, 113–14; by US Endangered Species Act, 114–15; of woodland caribou, 115–16
conservation values, conservation triage versus, 118–19
constant proportion strategy (CPS), 176–78, 177f

constant quota strategy (CQS): constant proportion strategy (CPS) compared to, 176–78, 177f; impact on populations, 172–75, 173f, 174f

consumer-resource dynamics of predation: bottom-up processes, 242–43, 242f; food chains, 236; trophic cascades, 237–42, 241f, 256n24

contest competition, 256n14

contraception, 58

controlled experiments, definition of, 256n26

Convention on International Trade in Endangered Species (CITES), 121, 168–69

cougar, predation on bighorn sheep, 39–40

count-based population viability analysis (PVA), 99–104, 101f, 103f, 252n7

cownose ray, trophic cascades and, 238–41

coyotes, compensatory mortality in, 195

crane, whooping: conservation ethics of, 45; extinction risk of, 110; log-transformed abundance of, 21–23, 23f; proportional growth of, 17, 18f, 19t

critically endangered species, 111–12, 111t

Crnokrak, Peter, 150, 151–52

crow, Hawaiian, extinction risk of, 95

cryptorchidism, in Florida panthers, 137

culling of elephants, Kruger National Park: adaptive management and, 62–63; building of arguments for, 51–54, 250n2; consequences of, 64–65; ecosystem health and, 59–61, 250n3, 250n5; estimates of abundance and, 50–51, 51f, 56–58, 57f; evaluation of premises in, 54–56; goals and methods of, 58–59; history of, 50–51, 51f; rationale for, 7–8; scientific uncertainty about, 61–62

cumulative distribution function for times to extinction, 253n19

D'Ancona, Umberto, 208

Darst, Catherine R., 108

DDT, 256n16

death rate, 14, 69

deep-sea fish: conservation ethics of, 45; density-independent dynamics in, 20–21, 24

deer, red, exploitation of, 195

deer, white-tailed, selective harvest of, 191–93, 204–5

delayed density dependence, 40–44, 41f, 43f, 44f

density-dependent dynamics: delayed, 40–44, 41f, 43f, 44f; environmental stochasticity, 35–37, 36f, 212–13, 255n4; logistic population growth, 29–31; negative, 25–26, 26f, 31f; nonlinear, 33–35, 34f; in northern gannets, 25–26, 26f; population growth, 27–31, 27f; positive, 37–40, 39f, 40f; prey growth, 225; strength of, 31–33, 31f, 32f; in structured populations, 251n6; time delays in, 42–43, 44f, 213

density-independent dynamics: of African bush elephants, 9–14, 11t, 13f, 249nn2–3; in decreasing populations, 20–21; definition of, 13; ecological processes for, 14–15; equations for, 15–17; exponential growth, 19–24, 23f; log-transformed abundance, 21–24, 23f; per capita growth rate, 12; projection matrices, 73–75, 73f; proportional growth, 9–19, 11t, 13f; recruitment, 10, 11t; in structured populations, 251n4; of whooping cranes, 17, 18f, 19t; of wolves, 17–19, 18f

deontology, 128, 250n4

desert topminnow, 150

desert tortoise, 108–10

diagrams, life-cycle, 70–72, 71f

Dickson, Paul, 50

Didinium nasutum, predation by, 215–17, 218f, 220, 255n5

diploid organisms, 142, 253n1

directional selection, 133
disassortative mating, 134
disruptive selection, 133
distance, as cause of nonrandom mating, 133–38, 134f
diversity, genetic. *See* genetic diversity
dogs, hunting with, 203–4
dolphins, resiliency to overexploitation, 197–98
dominant alleles, 136, 253n5
dominant eigenvalue, 78–79
dominant eigenvector, 80
drift, genetic. *See* genetic drift
Drosophila. See fruit flies (*Drosophila*), inbreeding depression in
dwarfism, in California condors, 137

eagle, bald, 256n16
eagle, golden, 230–31, 256n16
ecological processes, density-independent dynamics arising from, 14–15
ecosystem health, in conservation ethics, 59–61, 250n3, 250n5
effective population size: definition of, 157–60; estimation of, 160–61
eiders, 237
eigenvalue, dominant, 78–79
eigenvector, dominant, 80
elasticities, 80–82, 251n8
elephant, African bush: current and historic range of, 7–9, 8f; density-independent dynamics of, 9–14, 11t, 13f, 249nn2–3, 250nn6–7; estimates of abundance of, 51f, 56–58, 57f, 64, 249nn2–3; immigration of, 250n7; impact of exploitation on, 196; per capita growth rate of, 12, 250n9; recruitment rate of, 10, 11t; resiliency to overexploitation, 198. *See also* culling of elephants, Kruger National Park
elk: exploitation of, 199, 201–2; inbreeding depression in, 157–58; trophic cascades and, 238–41, 256n24
emergent properties, 5

endangered species, criteria for: Convention on International Trade in Endangered Species (CITES), 121, 168–69; geographic range, 123–26, 125f, 253n21; IUCN Red List Criteria, 95, 111–12, 111t, 123, 169–70; time and probability, 121–22, 122f; US Endangered Species Act (ESA), 121, 124–25, 125f. *See also* extinction risk
Endangered Species Act (ESA): conservation triage and, 114–15; criteria for endangered species under, 121, 124–25, 125f; extinction risk and, 95; polar bears and, 91–93, 92f
environmental governance, exploited populations and, 186–87
environmental stochasticity: in density-dependent populations, 35–37, 36f, 212–13, 251n2, 255n4; harvest strategies and, 179–87, 179f, 181f
equilibration, 80
equilibrium: in constant quota strategy (CQS), 175; equilibrium abundance, 78, 251n7; Hardy-Weinberg, 254n8, 254n15; lambda and, 78–79; population fragmentation and, 164; predation rate and, 230; stable, 31f, 80, 175–76; unstable, 175–76
Estes, John, 237
ethics: adaptive management, 62–63, 62f; argument analysis, 46–56, 63–64; confirmation bias and, 53–54; consequentialism, 250n4; conservation decision-making and, 62–63; definition of, 45–46; deontology, 250n4; ecosystem health, 59–61, 250n3, 250n5; exploitation, 203–4; goals and methods, 58–59; informing other ongoing cases with, 64–65; misperceptions about, 49–50; overabundance in, 56–58, 57f; precautionary principle in, 61; purpose of, 46; scientific uncertainty, 61–62; value of, 63–64

Excel-based simulations: constant quota strategy (CQS), 172–75, 174f; environmental stochasticity, 180–81; restricted proportion strategy, 184; threshold strategy, 184

exogenous sources of mortality, density-independent dynamics and, 14, 249n4

experimental designs, 256n26

exploited populations: assessing success or failure of, 201–3; benefits to human well-being, 169; best versus common practices for, 170, 204–5; bio-economics and environmental governance, 186–87; compensatory mortality in, 194–95, 194f; conflicting interests in, 170; environmental stochasticity in, 179–87, 179f, 181f; ethics of, 203–4; impacts beyond demography, 195–97, 255n9; legal versus illegal exploitation, 168; overconfidence, 200–201; overexploitation, 169–70, 197–200; scope of exploitation, 168–69; selective harvest, 191–93; structured populations, 187–93; threats to conservation, 169–70; unstructured populations, 170–79. *See also* harvest strategies

exploitive competition, 252n13

exponential growth: decreasing populations and, 20–21; distinguishing from other types of change, 21–23; equations for, 19–20; log-transformed abundance, 21–24, 23f

extinction risk: action plans for, 108–9; adaptive management for, 109–12, 120; conservation triage, 112–21; criteria for, 121–26; cumulative distribution function for times to extinction, 253n19; endangered species, criteria for, 121–26; hope and fatalism and, 126–29, 253n24; importance of, 94–95; inbreeding depression and, 153–57, 158f, 254n12; local extinctions, 1–2; multiple populations,

252n12; population viability analysis (PVA), 95–106, 108, 120–21, 156–57, 158f; societal willingness and ability to address, 111–12, 252n14; threats, actions, and outcomes, 108–9; universal minimum viable population size (MVP), 106–7, 253n19

Exxon Valdez oil spill, 238

fair chase, 203–4

Falkland Islands, southern elephant seals in, 158–59

fallacy of the undistributed middle, 48, 250n1

Fasta Åland, Glanville fritillary butterflies in, 154–55, 155f

fatalism, extinction risk and, 126–29

faux conservation triage, 119

field vole, 41

finite rate of increase, 78–79

fin whale, 197–98

fish: bass, 198–99; cod, 198, 200; cownose ray, 238–41; deep-sea, 20–21, 24, 45; desert topminnow, 150; exploitation of, 169, 196; mosquito fish, 157; onion-eye grenadier, 20–21, 22f, 24; range contraction of, 1–2; sea star, 237; sea urchin, 237–41; sunfish, 198–99; trout, 232

Fish and Wildlife Service (FWS), 118, 119

Florida panther: cryptorchidism in, 137; genetic rescue of, 165, 165f

foraging behavior: of alpine ibex, 37, 37f; density dependent, 25–26, 26f

fox(es), predation and, 44, 230–31

fox, arctic, predation by, 233–34

fox, San Joaquin kit, conservation triage for, 113–14

fox, swift, conservation triage for, 116

fragmentation, population, 162–64

frog, cascade, apparent competition and, 232

fruit flies (*Drosophila*), inbreeding depression in, 145, 153–54, 154f

heterozygotic individuals: allele frequencies and, 254n8; extinction risk and, 155, 155f; genetic diversity and, 131–32, 135f; heterozygotic advantage, 254n10

HexSim, 105

hippopotamus, pygmy, inbreeding in, 146–47, 147f

homozygotic individuals, 131–32, 135f, 142–43

hope, extinction risk and, 126–29, 253n24

Hörnfeldt, Birger, 41–42

Huffaker, Carl, 217–19, 218f, 220

human dimensions of wildlife conservation, 63

ibex, alpine: density-dependent dynamics of, 35–37, 36f; foraging behavior of, 37, 37f

ideal population, 159

illegal exploitation, 168

inbred mortality, 146–47, 147f

inbreeding coefficients, 138, 145, 153, 161, 163

inbreeding depression: definition of, 134–35, 136; difficulty in detecting, 151–53; examples of, 135–38, 135f; extinction and, 153–57, 254n12; genetic drift and, 142–45; genetic health and, 161–62; genetic rescue, 165–67, 165f; inbred mortality, 146–47, 147f; inbreeding coefficients, 135f, 138, 139f, 146, 163; in laboratory populations, 153–54, 154f; lethal alleles associated with, 136–38; minimum viable population size (MVP) and, 161–62; outbred mortality, 146–47, 147f; population size and, 142–45; skepticism of, 147–48; variable nature of, 151–53, 152f; in wild populations, 129, 130f, 148–51, 149f, 154–57, 155f, 158f; in zoo populations, 146–47, 147f

indexing variables, 251n3

individual-based population viability analysis (PVA), 105–6

inheritance patterns, 253n5

instantaneous per capita growth rate, 250n9

interference competition, 252n13

International Union for Conservation of Nature (IUCN), 24, 95, 106, 111–12, 111t, 123, 169–70

intraspecific competition: density dependence resulting from, 33, 37, 39–40; impact of, 14; lack of, 28–29

invertebrates: adzuki bean beetle, 218f, 219, 220; fruit fly, 145, 153–54, 154f; Glanville fritillary butterflies, 154–55, 155f; larch budmoth, 44; population viability analysis (PVA) of, 102; six-spotted mite, 217–19, 218f, 220; wasp, 218f, 219, 220

Isle Royale National Park moose population: age-specific vital rates for, 68–69, 68f, 70f; bottom-up predation processes, 242f; genetic rescue of wolves and, 166; per capita kill rate and, 3–4, 223–26, 224f, 226f; predation rate and, 226–30, 228f, 229f

Isle Royale National Park wolf population: bottom-up predation processes, 242–43, 242f; effective population size of, 160; genetic rescue of, 165–67, 165f; inbreeding depression in, 129, 130f, 139f, 148, 152–53; per capita kill rate and, 3–4, 59, 223–26, 224f, 226f; predation rate and, 226–30, 228f, 229f; wolf generations in, 254n16

ivory trade, 196

jackal, black, compensatory mortality in, 195

Jamieson, Ian G., 254n13

Kahneman, Daniel, 201

kelp forests, trophic cascades and, 237–41

Kenney, John, 164

Kenya, baboons in, 68f
kill rate, 222–25, 256n14
kite, snail, 34–35, 34f
Kluane National Park (Canada), red squirrels in, 133
Kruger National Park (South Africa), elephant population in: current and historic range of, 7–9, 8f; density-independent dynamics of, 9–14, 11t, 13f, 249nn2–3, 250nn6–7; estimates of abundance of, 56–58, 57f, 64, 249nn2–3; immigration of, 250n7; per capita growth rate of, 12, 250n9; recruitment rate of, 10, 11t. *See also* culling of elephants, Kruger National Park

laboratory populations, inbreeding depression in, 153–54, 154f
lambda, 78–79, 251n7; great tit/northern fulmar case example, 86f; Laysan albatross case example, 87–88; North Atlantic right whale case example, 90–91, 91f; polar bear case example, 93
larch budmoth, 44
lead poisoning, in Laysan albatross populations, 87–88
lemming(s), impact of predation on, 209–11, 212f, 233–34
Lennox, Robert J., 246
lethal alleles, 136–38, 157
life-cycle diagrams, 70–72, 71f; great tit/northern fulmar, 83–84, 83f; North Atlantic right whale case example, 89–90, 90f; polar bear case example, 91–93, 92f
Lindström, Jan, 33
lion(s): age-specific vital rates for, 68f; illegal trade in, 187; polyphagous predation and, 232
lion, mountain, predation by, 239
local extinctions, 1–2
logical fallacies, 47–49, 250n1
logistic population growth, 29–31
log-transformed abundance, 21–24, 23f

long-tailed skua, predation by, 233–34
Lotka, Alfred, 207
Lotka-Volterra equations: delicate nature of predator-prey coexistence in, 214–15; description of, 208–15, 208f; mechanisms within, 221–26; problem with, 219–20; result of, 212–14, 213f; shared credit for, 255n3. *See also* predation
lynx, predation by, 220–21, 221f

Macrourus berglax. See grenadier, onion-eye (*Macrourus berglax*)
Malthus, Thomas, 207
mammals. *See individual species*
Mandarte Island song sparrow population, inbreeding depression in, 148–51, 149f
marmot(s), density-dependent dynamics of, 38–39, 39f
marten, pine, predation by, 220
mating: assortative, 134; disassortative, 134; nonrandom, 133–38, 134f
matrices: dominant eigenvalue, 75–78; matrix multiplication, 75–78; projection, 72–75, 73f
maximum potential abundance, extinction risk and, 96
mechanisms, predation: density-dependent prey growth, 225; numerical response, 225–26, 226f; per capita kill rate, 222–25, 222f, 224f
mesocosms, 241–42
mesopredators, 245
metapopulations, extinction risk of, 102–4
Milk River Watershed, conservation triage in, 116–17
minimum viable population size (MVP): genetic diversity and, 161–62; universal, 106–7
mite, six-spotted, predation on, 217–19, 218f, 220
Mojave Desert, desert tortoises in, 108–10

SAFE index, 118, 253n18

saiga antelope: appearance of, 192f; selective harvest of, 192–93, 192f; 193f; threshold strategy for, 185

salmon, pink, allele frequency in, 132

San Joaquin kit fox, conservation triage for, 113–14

San Juan Channel, trophic cascades in, 238–41

scientific uncertainty, in conservation ethics, 61–62

scoters, 237

scramble competition, 256n14

seal, harbor, 150

seal, southern elephant, 158–59

sea otter, trophic cascades and, 237–41

sea star, predation on, 237

sea urchin, trophic cascades and, 237–41

secondary prey, 231

selachians, catch of, 208, 208f

selective harvest strategies: benefits to human well-being, 169; consequences of, 191–93; definition of, 187; sex-and stage-structured models, 187–91, 188f, 189f, 190f

senescence, 69

sensitivities, 80–82

sex-structured models, 187–91, 188f, 189f, 190f

sexual selection, 134

sharks, trophic cascades and, 238–41

sheep, bighorn: density-dependent dynamics of, 39–40, 40f; exploitation of, 195; genetic rescue of, 165

six-spotted mite, predation on, 217–19, 218f, 220

skua, long-tailed, predation by, 233–34

snakes, garter, apparent competition and, 232

snowshoe hare, predation on, 220–21, 221f

snowy owl, 233–34

socially informed adaptive management, 63

song repertoire, in Mandarte Island song sparrows, 149–50

South Africa, elephant population in. *See* Kruger National Park (South Africa), elephant population in

South America, bushmeat consumption in, 169

southern elephant seal, 158–59

space/distance, as cause of nonrandom mating, 133–38, 134f

sparrow, song, inbreeding depression in, 148–51, 149f

sparrow, white-throated, disassortative mating by, 134

spatial asynchrony, 103

sperm whale, 197–98

spiked speedwell (*Veronica spicata*), 154

squirrel, gray, 220

squirrel, red, 133

stabilizing selection, 132–33

stable age structure, 80

stable equilibrium, 31f, 175–76

stage-structured models, 71f, 104–5, 187–91, 188f, 189f, 190f

sterilization, 58

stoat(s), predation by, 210–11, 212f, 233–34

stochasticity, environmental. *See* environmental stochasticity

structured populations: conceptual foundation of, 68–69, 68f, 70f; definition of, 66–68; density-dependent, 251n6; density-independent, 251n4; dominant eigenvalue, 78; elasticities, 80–82, 251n8; equilibration, 80; great tit/northern fulmar, 82–86, 82f, 85f, 86f; Laysan albatross, 86–88, 88f; life-cycle diagrams for, 70–72, 71f; matrix multiplication, 75–78; North Atlantic right whale, 88–91, 90f, 91f; orca, 66–67, 67f; polar bear, 91–93, 92f; population vectors, 75; projection matrices for, 72–75, 73f; selective harvest strategies for, 187–93; sensitivities, 80–82; stage-structured, 71f, 104–5, 187–91, 188f, 189f, 190f